LABORATORY MANUAL

GENERAL CHEMISTRY 1113/1114

SOUTHERN METHODIST UNIVERSITY

Andrea S. Adams

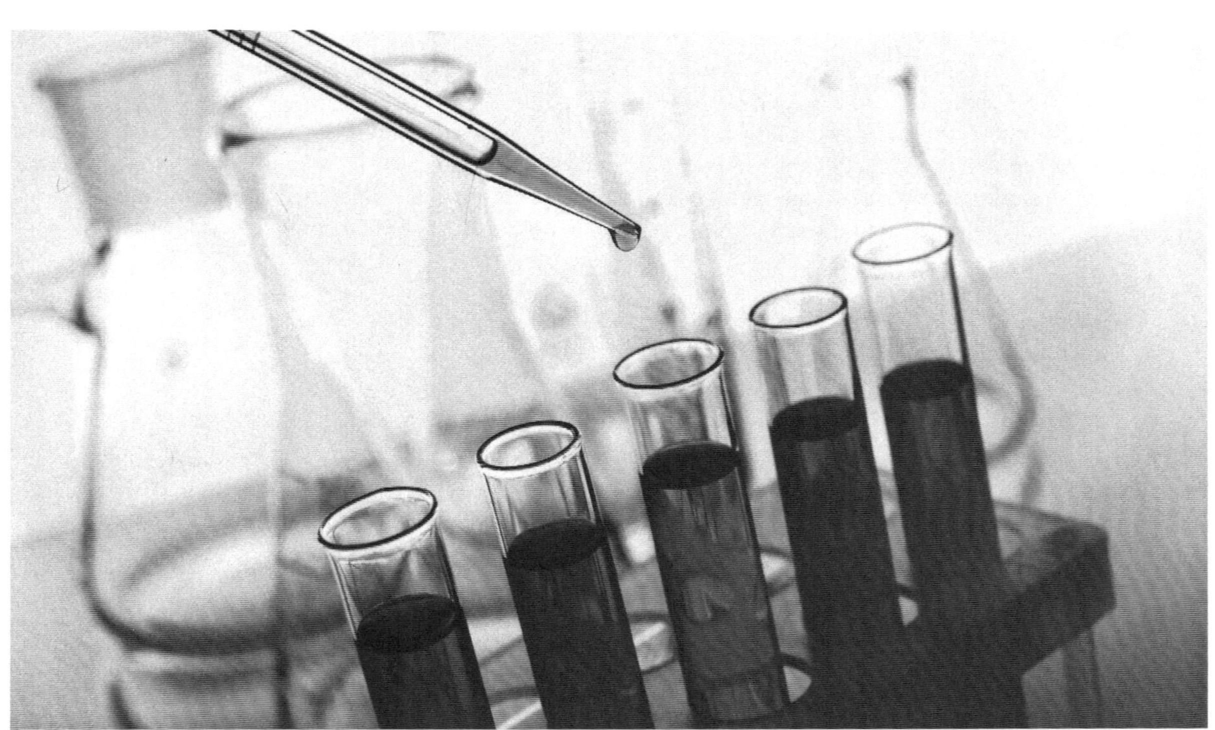

2018–2019

VAN-GRINER

General Chemistry 1113/1114

Laboratory Manual
Southern Methodist University
Andrea S. Adams
2018–2019

Printed in the United States of America
10 9 8 7 6 5 4 3 2
ISBN: 978-1-61740-640-9

Van-Griner Publishing
Cincinnati, Ohio
www.van-griner.com

CEO: Mike Griner
President: Dreis Van Landuyt
Project Manager: Hillary Lange
Customer Care Lead: Julie Reichert

Adams 640-9 Su18
208762-314174
Copyright © 2019

TABLE OF CONTENTS

Continued

APPENDICES

LABORATORY POLICIES AND SAFETY

If you are properly prepared, the experiments you will do in this course are not dangerous. However, if you do not understand the procedure to be followed, or if you are careless in performing the experiments, the chemistry laboratory can become hazardous. The following paragraphs discuss some laboratory safety rules which are observed in virtually every well run chemistry laboratory.

1. GOALS OF THIS COURSE
   UNDERSTAND WHAT YOU ARE DOING.
   The laboratory experiments in this manual have several purposes. First, they are intended to give a working association and understanding of some of the fundamental chemical principles presented in lecture. In addition, there is an opportunity to develop both proficiency and efficiency in many important laboratory techniques and in the acquisition and handling of experimental data. Finally, you will be asked to present and discuss your results in concise but complete reports covering the experiments performed.

2. USE THE REAGENTS WITH CARE.
   Make sure that you are using the right reagent. Read the label on a reagent bottle before you open it. Label any beakers or flasks in which you are storing reagent solutions. One of the most common and potentially dangerous mistakes is that of using the wrong reagent. There is no excuse for this. Also use only the quantities prescribed in the procedure.

3. YOUR PERSONAL PROTECTION IS IMPORTANT.
   - Obtain a pair of safety goggles or glasses and wear them at all times in the laboratory.
   - All students must wear a lab coat during the lab when chemicals are present.
   - No open toed shoes are allowed.
   - It is best not to wear contact lenses in the laboratory. Some contact lenses are degraded by chemical vapors. Others are permeable to some vapors and can trap them at the cornea.
   - Do not wear loose clothing that can drape over a Bunsen burner. Tie back long hair.
   - No food or drink is allowed in the laboratory at any time.

LABORATORY POLICIES AND SAFETY

4. <u>USE CAREFUL LABORATORY TECHNIQUE</u>.
   - Be careful in pouring liquids from reagent bottles.
   - Never insert a pipet or transfer pipet into a reagent bottle unless it is attached to the bottle for specific use of that solution.
   - Take only the amounts of reagents required for the experiment.
   - Do not return excess reagents to the bottles or solid chemical containers.
   - Use hot pads when handling hot materials.
   - Clean up your work space before you leave, sponging the desk thoroughly. Rinse all the glassware before checking out of the laboratory.
   - Wash your hands with soap and water before leaving the lab.
   - Wear gloves when necessary.

5. <u>BE AWARE OF THE LOCATION OF SAFETY EQUIPMENT</u>.
   Know where eyewashes, safety showers, and fire extinguishers are located.

6. <u>MATERIAL SAFETY DATA SHEETS</u>.
   A notebook containing Material Safety Data Sheets (MSDS) is available. These detail the properties of the different substances and solutions you will use in this course.

7. <u>DISPOSE OF EXCESS REAGENTS AND CHEMICAL WASTES CAREFULLY</u>.
   All chemical substances are potentially dangerous and can be detrimental to the environment. Some chemical wastes can be disposed of directly down the drain; others must be collected, stored, and disposed of using special methods. Follow instructions given by the Laboratory Instructor.

8. <u>LABORATORY POLICIES</u>
   All important laboratory policies are detailed in the syllabus provided/posted each semester.

Significant Figures Policy

All molar masses will be treated as exact numbers (an infinite number of significant figures) as long as they are obtained from the Lab Manual's periodic table and are written to the hundredths place.

Therefore, the number of significant figures in a problem will be dependent upon the number of significant figures in the given data. If the molar masses are not written to the hundredths place, the instructor can deduct points according to their discretion.

Examples
How many grams of iodine are formed from 5.4167 g of potassium permanganate? The balanced chemical reaction is as follows:

$$10 \, HI_{(aq)} + 2 \, KMnO_4{}_{(aq)} + 3 \, H_2SO_4{}_{(aq)} \rightarrow 5 \, I_2{}_{(aq)} + 2 \, MnSO_4{}_{(s)} + K_2SO_4{}_{(aq)} + 8 \, H_2O_{(l)}$$

$$5.4167 \text{ g KMnO}_4 \times \frac{1 \text{ mole KMnO}_4}{158.04 \text{ g KMnO}_4} \times \frac{5 \text{ mole I}_2}{2 \text{ mole KMnO}_4} \times \frac{253.80 \text{ g I}_2}{1 \text{ mole I}_2} = 21.747 \text{ g I}_2$$

- Since the molar masses are given to the hundredths place, base the answer's SF on the number of SF from the given data: the mass of the potassium permanganate (5 SF). The final answer would have 5 SF.

How many mL of water are formed? Refer to the above balanced equation. The density of water is 1.00 g/mL.

$$5.4167 \text{ g KMnO}_4 \times \frac{1 \text{ mole KMnO}_4}{158.04 \text{ g KMnO}_4} \times \frac{8 \text{ mole H}_2O}{2 \text{ mole KMnO}_4} \times \frac{18.02 \text{ g H}_2O}{1 \text{ mole H}_2O} \times \frac{1.00 \text{ mL}}{1.00 \text{ g}} = 2.47 \text{ mL H}_2O$$

- Since the molar masses are given to the hundredths place, base the answer's SF on the number of SF from the given data: the mass of the potassium permanganate (5 SF) and the density (3 SF). The final answer would have 3 SF.

How many grams of hydrogen react with 0.8256 g oxygen, based on the following equation:

$$2 \, H_2{}_{(g)} + O_2{}_{(g)} \rightarrow 2 \, H_2O_{(l)}$$

$$0.8256 \text{ g O}_2 \times \frac{1 \text{ mole O}_2}{32.00 \text{ g O}_2} \times \frac{2 \text{ mole H}_2}{1 \text{ mole O}_2} \times \frac{2.02 \text{ g H}_2}{1 \text{ mole H}_2} = 0.1042 \text{ g I}_2$$

- Since the molar masses are given to the hundredths place, base the answer's SF on the number of SF from the given data: the mass of the oxygen (4 SF). The final answer would have 4 SF.

If the following was written, where the molar masses are not given to the hundredth, the instructor could take off 1 point.

$$0.8256 \text{ g O}_2 \times \frac{1 \text{ mole O}_2}{32 \text{ g O}_2} \times \frac{2 \text{ mole H}_2}{1 \text{ mole O}_2} \times \frac{2 \text{ g H}_2}{1 \text{ mole H}_2} = 0.1032 \text{ g I}_2$$

v

Technique Lab – Chem 1113

Name________________________   Partner__________________________

Date________________________   Section__________________________

Part 1: Use of the buret, pipet, and analytical balance. Calculation of density and % error.
Important Points:
1.   USE UNITS AT ALL TIMES!!!
2.   The numbers in parenthesis are the number of significant figures.
3.   **DO NOT read volume by the beaker measurements! They are guidelines only: burets and pipets are much more accurate.**
4.   Record both the pipet and buret to the hundredth mL.

1. Normalize a 50 mL buret with DI water. After normalization, fill the buret with DI water between the 0 and 1 mL marks.	*Normalization* rinses a piece of equipment with the solution that will be delivered from it. This avoids contamination from substances previously in the buret. To normalize, add about 10 mL of the delivery solution to the buret. Remove the buret from the clamp and wet the inside of the buret by moving the buret at various angles. Drain the solution completely.
2. *Reading the buret*: Read the buret at the bottom of the meniscus, estimating to 0.01 mL	Initial volume: (2)
3. Fill a clean 250 mL beaker with approximately 100 mL of DI water.	
4. Weigh a clean and dry 50 mL beaker. Record the mass in the adjacent column. *Read the balance and record all four digits after the decimal*. Bring this data sheet and a writing instrument to the balance room.	Mass: (all)
5. Normalize a 10.00 mL pipet by drawing up solution slightly over the mark above the barrel and discarding it to waste. Pipet a 10.00 mL aliquot of DI water from the 250 mL beaker and transfer this to the 50 mL beaker. Record the volume delivered. The volume of a pipet should be read to the nearest 0.01 mL. Since a pipet delivers an exact amount of liquid, when the bottom of the meniscus is on the mark, the volume delivered is 10.00 mL.	*To pipet*: Gently insert a pipet into the pipet pump (autopipetter). Turn the wheel of the pipet pump with your thumb to draw up liquid until the bottom of meniscus is at the mark on the neck of the pipet. Depress the white bar on the side of the pipet pump to dispense. Touch the pipet tip to the beaker when done. **** Never let liquid enter the pipet pump. Always keep the level of the pipet pump above the level of the pipet. **** Volume delivered: (4)

6. Weigh the beaker and water. Record the mass. Do not discard the water.	Mass: (all)
7. Deliver approximately 5 mL of the water in the buret to the 50 mL beaker. Record the volume delivered. The volume of a burette should be read to the nearest 0.1 of a mL, and then estimated to the nearest 0.01 of a mL. (Readings should be in the form X.XX mL.) Read the buret at the bottom of the meniscus.	Initial volume (from step 2): Final volume: Volume delivered: (3)
8. Weigh the beaker and water. Record the mass. Do not discard the water.	Mass: (all)
9. Pipet a 10.00 mL aliquot of DI water from the 250 mL beaker and transfer this to the 50 mL beaker. Record the volume delivered.	Volume delivered: (4)
10. Weigh the beaker and water. Record the mass. Do not discard the water.	Mass: (all)
11. Deliver approximately 10 mL of the water in the buret to the 50 mL beaker. Record the volume delivered.	Initial volume: Final volume: Volume delivered: (4)
12. Weigh the beaker and water. Record the mass.	Mass: (all)
13. Calculate the density of the water. Density $=\ \dfrac{\text{mass}}{\text{volume}}$ where mass of water = #12 – #4 and volume = #5+ #7 + #9 + #11	Density: (be sure to include units) (4)
14. Water typically has a density of 1.0000 g/mL. Calculate a % error of the experimental value vs the theoretical value: % error $=\ \dfrac{\text{experimental value - theoretical value}}{\text{theoretical value}} \times 100$	% error: (4)

Note: A positive % error means the experimental value is larger than the theoretical value. A negative % error means the experimental value is smaller than the theoretical value. It gives a magnitude of the error, but does not indicate the type (systematic or random) of error.

TECHNIQUE LAB: CHEM 1113

Name_________________________   Partner___________________________

Date_________________________   Section___________________________

Part 2: Use of the thermometer, Bunsen burner, and reading the temperature gauge in the lab.
Calculation of °C and K and % error. USE UNITS!!!

1. Using a 0 – 100 °C thermometer, determine the temperature of the air in the lab. *This temperature should be read to the nearest 0.1 degree.*	Temperature: (3)
2. Place a Bunsen burner (not lit) under the ring clamp/gauze that is attached to the ring stand. Adjust the height of the ring clamp so that it is 2-3 inches above the top of the Bunsen burner.	
3. Fill a 250 mL beaker approximately halfway with DI water and place it on the ring clamp/gauze set up.	
4. Insert the thermometer into the thermometer clamp on the ring stand and lower it into the water, making sure the thermometer is not touching the bottom of the beaker.	Temperature: (3)
5. Light the Bunsen burner and heat the water until it is boiling.	
6. Read the temperature of the boiling water.	Temperature: (tenths)
7. Convert the thermometer (recorded in step #6) reading to °F. $$\left(\frac{9}{5}\right)°C + 32 = °F$$	Temperature: (3 if T < 100 or 4 if T ≥ 100)
8. Assuming that the theoretical value of boiling water is 212.0 °F, calculate a % error of the thermometer temperature. $$\% \text{ error} = \frac{\text{experimental value - theoretical value}}{\text{theoretical value}} \times 100$$	% Error: (2)
9. Based on the % error, is the thermometer temperature reliable? (Should be ± 5%)	
10. Convert the boiling point of water obtained in #6 to K. $$K = °C + 273.15$$	Temperature at boiling, K: (tenths)

Part 3: Reading the barometer and converting units.

Read the red scale (mm Hg) of the barometer in the lab. Since each subdivision of this scale is 1 mm Hg, estimate to the tenths place.

Record the value: ___________________ mm of Hg

Number of significant figures in this value: _____________

Many units are used for pressure. Using the following conversion factors and dimensional analysis, convert the lab barometer reading to the necessary units.

1 inch Hg = 25.4 mm Hg

1 torr = 1 mm Hg

1 atm = 760 mm Hg

1 atm = 101,325 Pa

<table>
<tr><td>Important Point! Read Carefully!
These values are exact numbers. Exact numbers are numbers obtained from definitions (conversion factors) or by counting the number of objects in a group (such as "12" are in a dozen). Exact numbers have an infinite number of significant figures. Therefore, the number of significant figures in a converted answer below is dependent upon the number of significant figures in the barometer reading.</td></tr>
</table>

<u>SHOW ALL WORK USING DIMENSIONAL ANALYSIS!</u>

Barometric pressure, torr: ______________________

Barometric pressure, inches of Hg: ______________________

Barometric pressure, atm: ______________________

Barometric pressure, Pa: ______________________

<u>Laboratory Basics Homework</u>

<u>This homework is 8 pages.</u>

<u>OBSERVE SIGNIFICANT FIGURES ON ALL CALCULATIONS; SHOW ALL WORK</u>

1. *Identify the following pieces of equipment, writing the name on the line.*

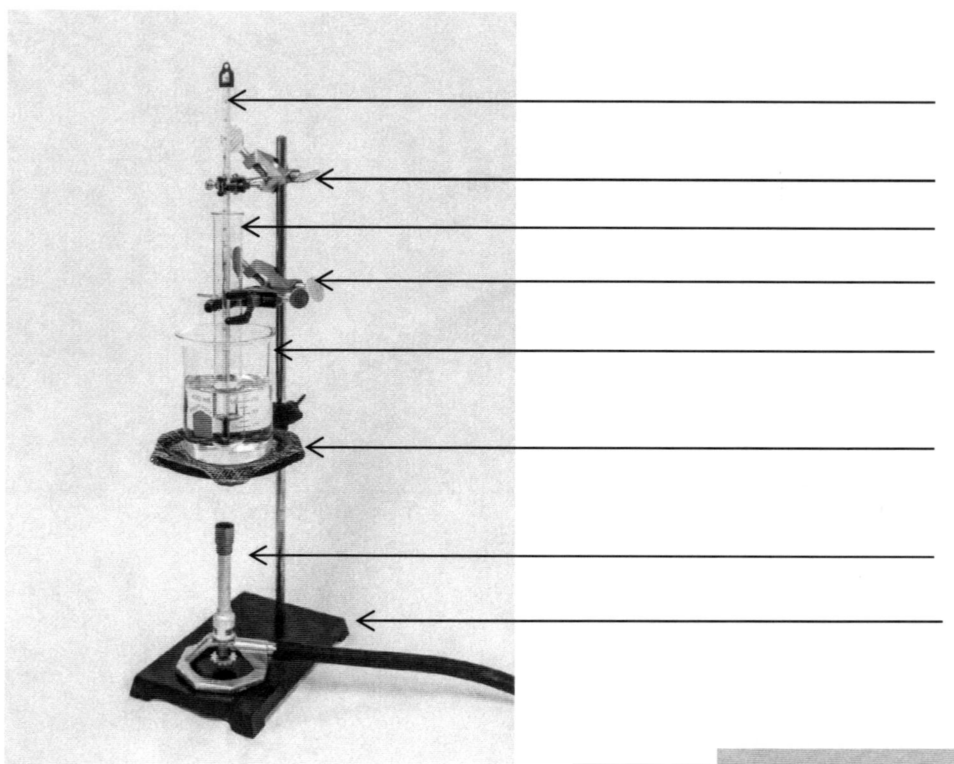

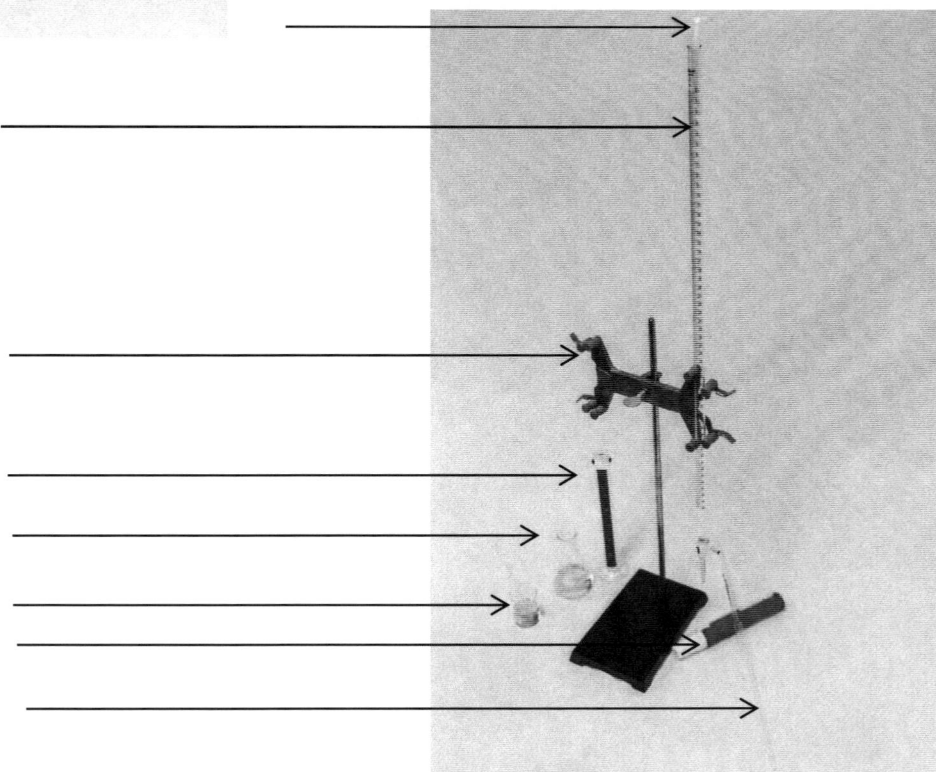

2. *Significant Figures.* In this section, although an answer may look correct, if the number of significant figures in the answer is incorrect, no credit will be given for the problem.

   a. List the number of significant figures in the following numbers

4057 _______	0.57 _______	0.7500 _______
1020 _______	0.00391 _______	1020. _______
1020.0 _______	50007 _______	0.056340 _______

   b. Calculate the following

   $14.7 + 0.62 =$ _______________ $0.6257 - 0.031 =$ _______________

   $235.67 + 44.980 - 5.2 =$ _______ $67.9 \times 57.224 =$ _______________

   $0.0045 \times 1.882 =$ _______________ $5959.2 \div 24.4 =$ _______________

   $1.06 \div 0.02 =$ _______________ $\dfrac{(14.55 - 9.2)^2 - 0.25}{0.00456 \times 1.9973} =$ _______________

   c. An electrolysis cell deposits copper at a rate of 3.069 g of copper per minute. How many grams of copper will be deposited in 2.0 hours?

   d. Isopropyl alcohol evaporated from a beaker at the rate of 1.057 mL/min. How many seconds would it take for 3.70 mL of isopropyl alcohol to be lost?

   e. Given that 6.02×10^{23} ethanol molecules have a mass of 46.04 g, calculate the number of ethanol molecules in a film of ethanol that is 3.08 mm by 10.0 cm by 0.0015 mm. Assume the density of ethanol is 0.790 g/mL.
   Strategy: First calculate the volume in cm³, convert to g, then convert g to molecules.

LABORATORY BASICS HOMEWORK

f. The density of an irregular shaped object can be found in the following way:
    1) Weigh the object to obtain the mass;
    2) Put some water in a graduated cylinder and read the volume;
    3) Place the object in the cylinder and reread the volume;
    4) The volume of the object is the difference in volume reading in steps 2 and 3.

Complete the missing values using the proper number of significant figures. <u>Include units!</u>

a. mass of object   18.9262 g
b. volume before object added   8.97 mL
c. volume after object added   16.48 mL

d. volume of object   ____________

e. density of object   ____________

3. *Percent Error.* Observe significant figure rules.

$$\% \text{ error} = \frac{\text{experimental value - theoretical value}}{\text{theoretical value}} \times 100$$

a. Calculate the percent error if a thermometer reads 76.25 °C, and a reference thermometer reads 74.96 °C.

b. The heat of neutralization in an experiment was determined to be −58.7 kJ/mol. The literature value for this heat of neutralization is −56.2 kJ/mol. Calculate the percent error.

4. *Interpolation.* Using the Appendix, correctly determine the vapor pressure of water at the following temperatures. Carefully follow the interpolation instructions in the text box on the page. Show all work!!! Observe significant figure rules.

	<u>atm</u>	<u>mm Hg</u>	<u>Work</u>
24.3 °C	________	________	
27.6 °C	________	________	

7

5. *Equipment Reading.*

To correctly read a piece of equipment, first determine the magnitude of the smallest division (markings). Then, estimate the reading in the next smallest order of magnitude. For example, if a thermometer is marked in one-degree increments, the reading would be estimated to the tenth of a degree. If it is marked in tenth of a degree increments, the reading would be estimated to the hundredth of a degree. Read at the bottom (or top, if curved upward) of the meniscus. Correctly read (*with abbreviations for units*) the following:

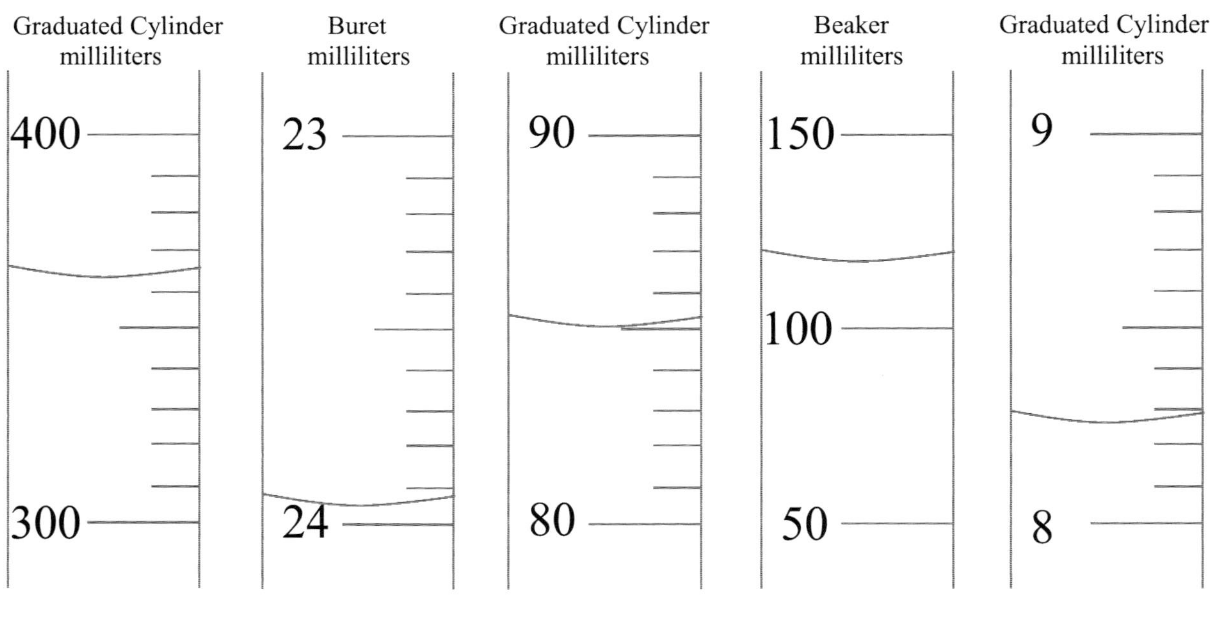

LABORATORY BASICS HOMEWORK

Read the following laboratory equipment, <u>with abbreviations for units</u>:

Buret
(milliliters)

Buret
(milliliters)

Pressure Gauge (read the middle scale, mm Hg)

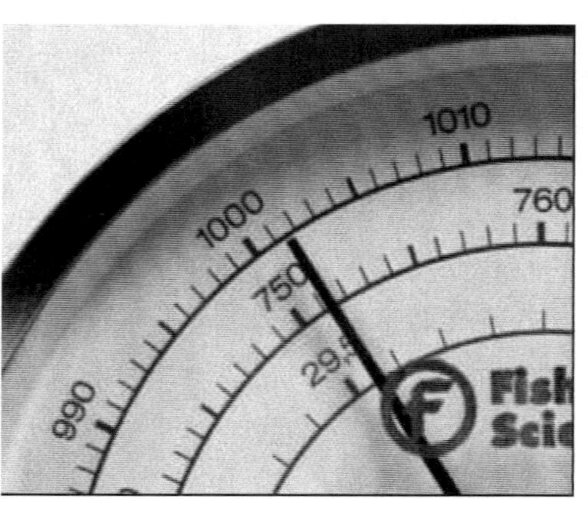

_______________   _______________   ___________________________

Thermometer
(degrees Celsius)

Thermometer
(degrees Celsius)

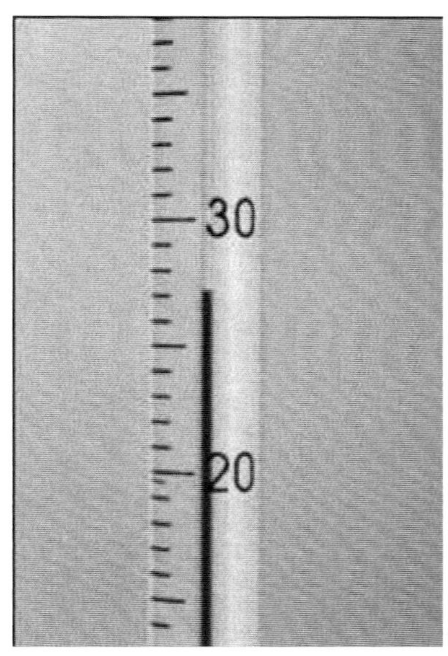

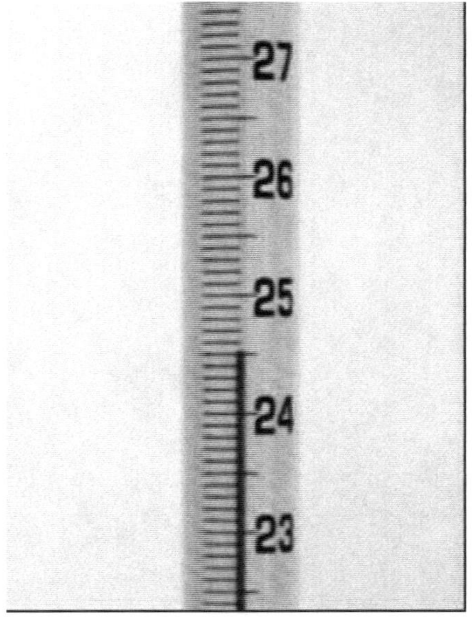

_______________________   _______________________

6. *Graph Construction.*

Using the graphing instructions in the Appendix, construct a graph of the following data. Connect the points smoothly. This is a titration of 50.0 mL of 0.056 M oxalic acid (a diprotic acid) with 0.400 M NaOH.

mL NaOH Added	pH		mL NaOH Added	pH		mL NaOH Added	pH
0.00	1.14		12.60	5.29		23.50	9.67
1.10	1.18		13.20	5.54		24.70	10.50
2.50	1.28		13.80	5.77		26.20	10.91
3.70	1.41		14.70	6.02		27.40	11.15
5.20	1.63		16.00	6.29		28.80	11.36
7.30	1.98		17.50	6.58		30.40	11.61
10.20	2.80		19.20	6.92		32.10	11.87
10.90	3.40		20.90	7.31		33.40	12.12
11.80	4.70		22.80	8.26			

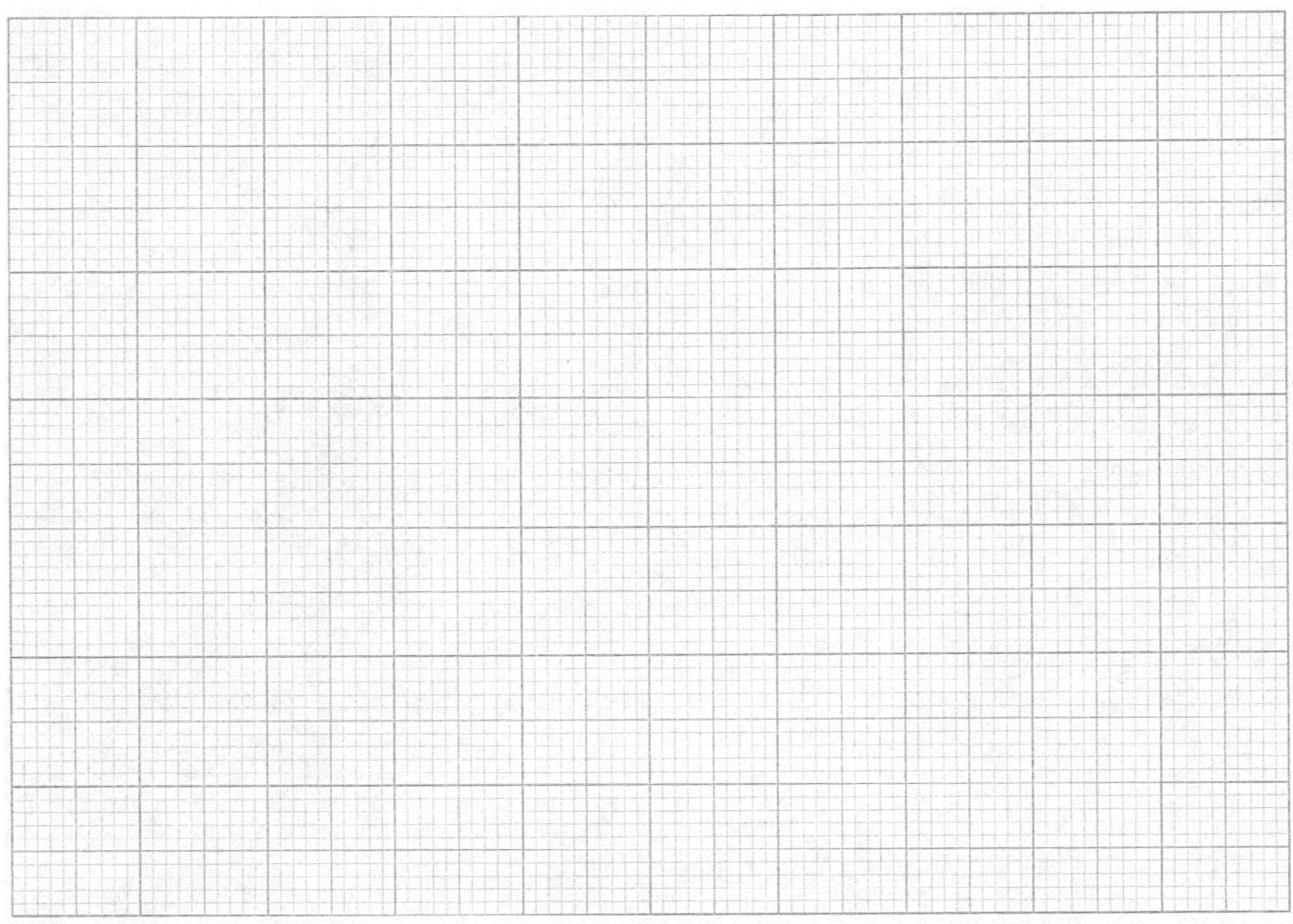

LABORATORY BASICS HOMEWORK

7. *Convert the following units.* <u>For parts b-d, dimensional analysis MUST be used to receive credit.</u> Some conversion factors and formulas can be found in the Appendix. Show all work and box all answers. Observe significant figures rules.

   a. <u>Temperature</u>

       75.2 °C to K

       78.0 °F to °C

       23.62 °C to °F

   b. <u>Pressure</u>

       Example: 29.17 inches of Hg to mm Hg $29.17 \text{ in Hg} \times \dfrac{760 \text{ mm Hg}}{29.92 \text{ in Hg}} = 740.9 \text{ mm Hg}$

       29.17 inches of Hg to Pa

       29.17 inches Hg to atm

   c. <u>Moles, mass, and particles</u>

       Example: 0.256 moles N_2 to g N_2 $0.256 \text{ moles } N_2 \times \dfrac{28.02 \text{ g } N_2}{1 \text{ mole } N_2} = 7.17 \text{ g } N_2$

       53.2895 g water to moles water

       Number of atoms of sulfur in 0.0218 kg of $Na_2S_2O_3$.

11

d. <u>Density</u> (Note: 1 mL = 1 cm^3)

The density of iron is 7.86 g/cm^3. Calculate the density in kg/m^3.

At 1 atm, the density of air is 1.225 kg/m^3. Calculate the density in g/L.

0.1265 kg gallium to mL gallium (the density of gallium is 5.91 g/cm^3)

7. *Miscellaneous*
    a. If too much solid unknown was put on a weighing paper on the balance, what should you do, and where do you put any extra solid?

    a. If too much solution was taken, what do you do with the excess solution?

    b. What does tare mean, when do you do tare something, and how do you do that?

    c. What do you do with your buret when done with the experiment?

EXPERIMENT 1
DENSITY OF LIQUIDS AND SOLIDS

The densities of pure 2-propanol, pure water, and two 2-propanol/water solutions will be measured, and the mass percent of water in each is determined. A graph of the density of each sample versus its mass percent of water present will be constructed. The density of an unknown 2-propanol/water solution will be measured, and the graph will then be used to determine the composition of the unknown. The density and identity of an unknown solid will be determined.

Key Chemical Reactions: None

Key Mathematical Equations:

$$\text{density} = \frac{\text{mass}}{\text{volume}}$$

$$\text{mass percent of water} = \frac{\text{grams of water in the solution}}{\text{total grams of solution}} \times 100$$

$$\% \text{ error} = \frac{\text{experimental value - theoretical value}}{\text{theoretical value}} \times 100$$

Discussion

Density is the ratio of the mass to the volume of a substance.

$$\text{density} = \frac{\text{mass}}{\text{volume}}$$

The density of a pure substance is a characteristic physical property, which can be useful in identifying a substance. Density is an intensive property because it does not depend on the amount of the substance present. The SI unit of density is kg/m^3. However, in most cases it is more convenient to express density in g/mL or in g/cm^3. Since 1 mL = 1 cm^3, density values of g/mL and g/cm^3 are numerically equal. Density varies with temperature, so the temperature at which an experiment is performed should be recorded.

If two of the three values in the density equation are known, the third value can be determined. The use of dimensional analysis is the best method to solve density problems. For example, a sample of pure silver (density of 10.5 g/mL) was found to have a mass of 2.0781 grams. Determine the volume of the silver sample.

$$2.0781 \text{ g Ag} \times \frac{\text{mL}}{10.5 \text{ g Ag}} = 0.198 \text{ mL}$$

When two (or more) pure liquids are mixed, the density of the resulting solution depends on the densities of the pure liquids, the relative amounts of the liquids in the mixture, and the temperature. If a solution made by mixing liquids behaves ideally (known as an ideal solution), the volumes of the liquids will be additive and the density of a particular composition can be calculated from the densities of the pure liquids.

However, the density behavior of real (non-ideal) solutions is more complicated. When liquids of non-ideal solutions are mixed, the volumes are not necessarily additive. The volume of the resulting solution can be either greater or less than the sum of the volumes of the pure liquids depending on the nature of the interaction between the liquids. Therefore, the density versus composition relation must be determined experimentally, as is done in this experiment.

Calculations and Examples

Ideal Solution Example
In an experiment, a sample of gasohol (a gasoline/ethyl alcohol blend) is made and contains 10.00% ethyl alcohol (ethanol) by mass. The solution is determined to behave ideally. The densities of gasoline and ethanol at 20 °C were determined to be 0.7512 g/mL and 0.7389 g/mL, respectively. Determine the density of the blend.

First, assume that 100.0 grams of the solution will be made. This means that it will be comprised of 10.00 grams of ethanol and 90.00 grams of gasoline. Next, calculate the volume of each component.

$$10.00 \text{ g ethanol } \times \frac{\text{mL}}{0.7389 \text{ g ethanol}} = 13.53 \text{ mL ethanol}$$

$$90.00 \text{ g gasoline } \times \frac{\text{mL}}{0.7512 \text{ g gasoline}} = 119.8 \text{ mL gasoline}$$

Since the solution behaves ideally, the volumes are additive. The total volume of the solution is 13.53 mL + 119.8 mL = 133.3 mL.

The density equation is then used to determine the density.

$$\text{density} = \frac{100.0 \text{ g}}{133.3 \text{ mL}} = 0.7502 \text{ g/mL}$$

Nonideal Solution Example
When dealing with nonideal solutions (this experiment), the volumes are not additive, and the densities of the pure liquids are used to determine the mass of each component in the solution. For example, when 10.00 mL of ethanol and 20.00 mL of water are mixed:

$$10.00 \text{ mL ethanol } \times \frac{0.7389 \text{ g ethanol}}{\text{mL}} = 7.39 \text{ g ethanol} \qquad 20.00 \text{ mL water } \times \frac{0.9917 \text{ g water}}{\text{mL}} = 19.83 \text{ g water}$$

EXPERIMENT 1   DENSITY OF LIQUIDS AND SOLIDS

The mass percent of water present in a solution can be calculated:

$$\text{mass percent of water} = \frac{\text{grams of water in the solution}}{\text{total grams of solution}} \times 100$$

$$\text{mass \% water} = \frac{19.83 \text{ g water}}{7.39 \text{ g ethanol} + 19.83 \text{ g water}} \times 100 = 72.85\,\%$$

The graphical relationship of the density versus the mass percent water for various solutions of the same components enables the density of a specific composition to be determined.

<u>Chemicals</u>
    2-propanol C_3H_7OH
    Various solids

Solid	Density at 20 °C		Solid	Density at 20 °C
Aluminum	2.70 g/mL		Iron	7.87 g/mL
Brass	8.00 g/mL		Lead	11.35 g/mL
Copper	8.92 g/mL		Polyvinyl chloride	1.14 g/mL
Glass	2.60 g/mL		Steel	7.60 g/mL
Graphite	2.25 g/mL		Zinc	7.14 g/mL

<u>Waste</u>
    All waste can be disposed down the laboratory sink.

<u>Procedure</u>
<u>Part 1: Determination of Pure Liquid Densities</u>
1. Obtain about 80 mL of 2-propanol in a 100 mL beaker.

2. Normalize two burets, one with deionized (DI) water and one with 2-propanol. Use only 5 –10 mL of each liquid, making sure that the liquid wets all parts of the buret that could come in contact with that liquid when the buret is filled. Drain the normalizing liquid to the sink.

3. Fill the burets with the solution they will deliver and then drain them below the 0.00 mL mark, making sure there are no bubbles in the buret tip. Read each buret to the nearest 0.01 mL and record these initial readings on the data sheet.

4. Weigh a clean, dry, and stoppered 50 mL flask. Make sure that the outside of the flask is also dry before it is weighed. Record all digits displayed on the balance.

5. Dispense approximately 10 mL of DI water from the buret into the weighed flask and restopper. Carefully record the exact volume before and after the liquid is delivered.
    The volumes of the burets should be read to the nearest 0.01 mL. It is not necessary to drain the exact volume listed above; however, the exact volume delivered should be recorded. For example, if ~10 mL of water should be delivered, it is acceptable to drain 9.98 mL or 10.05 mL, as long as these numbers are recorded in the data sheet.

15

6. Reweigh the flask, stopper, and liquid.

7. With a new clean, dry, and stoppered 50 mL flask, repeat steps 4 and 5 using pure 2-propanol.

8. Complete the density calculations of each pure liquid on the data sheet.

Part 2: Preparation of Solutions
1. Prepare the following solutions (draining directly from the buret to the test tube) in clean and dry large test tubes, recording the exact volume before and after each liquid is delivered. Make sure there is enough liquid in each buret before each addition.

       1.   15 mL H_2O + 30 mL 2-propanol
       2.   30 mL H_2O + 15 mL 2-propanol

2. Stopper and mix the solutions thoroughly by gently shaking the test tubes in a back-and forth motion. If only one stopper is provided, thoroughly dry it before placing on a new test tube. The solutions can also be mixed with a stirring rod. Thoroughly dry it before placing in a new test tube.

3. Complete the calculations in Part 2 of the data sheet using the densities determined in Part 1.

Part 3: Determination of Solution Densities
1. Weigh a clean, dry, and stoppered 50 mL flask. Make sure that the outside of the flask is also dry before it is weighed. Record all digits.

2. Normalize a 10.00 mL pipet with solution 1 (prepared in Part 2).

3. Pipet a 10.00 mL aliquot (a measured part of a whole) of the solution into the flask. The autopipetter unit should never be at or below the level of the pipet. Never let liquid enter the autopipetter.

4. Stopper the flask, reweigh it, and calculate the density of the solution.

5. Repeat steps 1 through 4 with solution 2, using a separate clean and dry flask. Be sure to normalize the pipet with the new solution.

6. Complete the calculations in Part 3 of the data sheet.

Graphing of Data
Plot a graph of the density versus the mass percent of water in the solutions. Refer to the following points and the Appendix for proper graphing techniques.
1. Orient the graph in the landscape arrangement. Use the y axis (shorter axis) for density and the x axis (longer axis) for mass percent water.
2. In setting the coordinate scales, the range of the mass percent of water will be from zero (pure 2-propanol) to 100.0 percent (pure water). Use an appropriate scale (perhaps starting at 0.7500 g/mL) for density.

3. Label axes with units where applicable and be sure the graph has an appropriate title. (The title is always dependent variable vs. independent variable, or the y label vs. the x label.)
4. Plot the densities and mass percent water of the two solutions, as well as the two pure liquids. There should be four points on the graph.
5. Draw a smooth curve through the points. The densities of the solutions should increase as the mass percent of water increases. Since the solutions are not ideal, the density vs. mass percent relationship is not linear.

Part 4: Determination of the Mass Percent of Water in an Unknown
1. Weigh a clean, dry, and stoppered 50 mL flask. Make sure that the outside of the flask is also dry before it is weighed. Record all digits.

2. Using the autodispenser, dispense 10.0 mL of the unknown solution to the flask.

3. Stopper the flask, reweigh it, and calculate the density of the solution.

4. From the graph of density versus mass percent of water constructed in Part 3, determine the mass percent of water in the unknown solution using the calculated density.

Part 5: Determination of the Density of a Solid
1. Weigh and record the entire amount of the unknown solid.

2. Add water to a 25 mL graduated cylinder so that the water comes between the 10 and 15 mL marks. Record the volume, estimating to the nearest 0.1 mL.

3. Carefully add the entire amount of unknown solid to the graduated cylinder and record the new volume. Calculate the density of the solid. Comparing to the data provided, determine the identity of the solid.

4. Remove the solid from the graduated cylinder and completely dry it before returning the solid to its container.

Conclusion:
In a short paragraph written in a technical manner, summarize in two or three sentences the purpose of the lab. Next, in sentence form, detail <u>relevant</u> data from the laboratory report sheet, in logical order, including units where appropriate. Be sure to include unknown numbers. It is helpful to underline all numerical data.

EXPERIMENT 1   DENSITY OF LIQUIDS AND SOLIDS

EXPERIMENT 1   DENSITY OF LIQUIDS AND SOLIDS

LABORATORY REPORT SHEET

Name_______________________________   Room Temperature_________°C

1. DENSITIES OF PURE LIQUIDS

INCLUDE ALL UNITS AND SIG FIGS!

	Water	2-propanol
Initial buret reading		
Final buret reading		
Volume of sample		
Mass of flask + stopper		
Mass of sample + flask + stopper		
Mass of sample		
Density		

2. SOLUTION PREPARATION

15 mL water and 30 mL 2-propanol	Water	2-propanol
Initial buret reading		
Final buret reading		
Volume of water, 2-propanol		
Mass of water, 2-propanol		
Total mass of solution		
Mass % water		

30 mL water and 15 mL 2-propanol	Water	2-propanol
Initial buret reading		
Final buret reading		
Volume of water, 2-propanol		
Mass of water, 2-propanol		
Total mass of solution		
Mass % water		

EXPERIMENT 1   DENSITY OF LIQUIDS AND SOLIDS

3. DENSITY OF SOLUTIONS

> **INCLUDE ALL UNITS AND SIG FIGS!**

	15 mL H_2O 30 mL 2-propanol	30 mL H_2O 15 mL 2-propanol
Mass of flask + stopper		
Mass of flask + stopper + sample		
Mass of sample		
Volume of sample		
Density of sample, g/mL		

4. UNKNOWN SOLUTION DENSITY AND MASS % WATER

Unknown #________

Mass of flask + stopper	
Mass of sample + flask + stopper	
Mass of solution	
Volume of solution	
Density of solution	
Mass % of water in the solution	

5. UNKNOWN SOLID DENSITY AND IDENTITY   Unknown #________

Mass of solid	
Initial volume of water	
Final volume of water	
Volume of solid	
Density of solid: ______________ Identity of unknown solid:________________	

Error Analysis /Questions:
1. Calculate the percent error of the density of the unknown solid compared to the theoretical value (listed in the chemicals section).

2. Complete the following statement relating an error to density and mass % water.
   Example: If there were air bubbles in the tip of the buret delivering 2-propanol, less 2-propanol would be delivered than recorded, and the density would appear greater for the same mass % water.

   If the flask containing the unknown solution was not stoppered quickly, some

   of the solution would evaporate. The mass recorded for the sample will be _________
   higher/lower

   than what was delivered and recorded, and the density of the unknown would appear

   ________________.   The mass % water in the unknown would appear _____________.
   higher/lower   higher/lower

Name _______________________________

<u>HOMEWORK EXERCISES</u>

1. Describe the procedure of normalization for a buret.

2. Describe the procedure of normalization for a pipet.

3. In an experiment at 20 °C, the density of methanol is found to be 0.7915 g/mL and that of water is 0.9917 g/mL. A solution is prepared by mixing 19.75 mL of the methanol with 18.25 mL of water. Calculate:

a. The mass of water in the solution

b. The mass of methanol in the solution

c. The mass percent of water in the solution

d. The mass percent of methanol in the solution

e. A 10.00 mL aliquot of this solution is pipetted and found to have a mass of 9.2693 g. Calculate the density of the solution.

f. Compare the calculated answer for the density with the literature value of 0.9072 g/mL using the % error calculation. Refer to the Appendix for the formula.

g. Using the value obtained above in (e), calculate the density in kg/m^3.

Continued on back

4. In an experiment analyzing a hand sanitizer at 20°C, the density of ethanol, isopropanol, and water were determined to be 0.7839 g/mL, 0.7855 g/mL, and 0.9928 g/mL, respectively. What would be the density of an ideal solution containing 62.00% by mass of ethanol and 10.00% by mass of isopropanol?

5. A small chunk of iron was determined to have a mass of 8.2978 grams. It was placed in a graduated cylinder originally containing 7.70 mL of water and the water level rose to 8.76 mL. Calculate the density of iron.

6. The following data was obtained for six solutions of sulfuric acid and water:

mass % water	100.0	90.00	70.00	50.00	20.00	0.00
density (g/mL)	0.9956	1.0661	1.2191	1.3952	1.7272	1.8305

   a. Plot a graph of density versus mass percent water using the above data on graph paper, which can be found at the back of this manual. Refer to the Appendix for graphing techniques. Orient the graph landscape, label all axes and title (with units), use the entire paper, and draw a smooth line connecting the points.

   b. 50.00 mL of a sulfuric acid and water solution has a mass of 82.4352 g. From the graph plotted in (a), what is the mass percent of water in the solution?

EXPERIMENT 2
THE COPPER CYCLE

A known mass of copper metal is dissolved and taken through a series of chemical reactions forming different copper compounds, finally ending up as copper metal again. The percent of the copper recovered will be calculated. This experiment will give practice in the quantitative transfer of both solutions and solids, as well as practice with gravity and suction filtration.

Key Chemical Reactions:

$$Cu_{(s)} + 4\,HNO_{3(aq)} \rightarrow Cu(NO_3)_{2(aq)} + 2\,NO_{2(g)} + 2\,H_2O_{(l)}$$

$$Cu(NO_3)_{2(aq)} + 2\,NaOH_{(aq)} \rightarrow Cu(OH)_{2(s)} + 2\,NaNO_{3(aq)}$$

$$Cu(OH)_{2(s)} \rightarrow CuO_{(s)} + H_2O_{(l)}$$

$$CuO_{(s)} + 2\,HCl_{(aq)} \rightarrow CuCl_{2(aq)} + H_2O_{(l)}$$

$$CuCl_{2(aq)} + Zn_{(s)} \rightarrow ZnCl_{2(aq)} + Cu_{(s)}$$

$$Zn_{(s)} + 2\,HCl_{(aq)} \rightarrow ZnCl_{2(aq)} + H_{2(g)}$$

Key Mathematical Equations:

$$\text{percent recovery} = \frac{\text{grams of copper recovered}}{\text{grams of initial sample}} \times 100$$

dimensional analysis:

$$1.50\ \text{g Cu} \times \frac{1\ \text{mole Cu}}{63.55\ \text{g Cu}} \times \frac{4\ \text{moles HNO}_3}{1\ \text{mole Cu}} \times \frac{63.02\ \text{g HNO}_3}{1\ \text{mole HNO}_3} = 5.95\ \text{g HNO}_3$$

EXPERIMENT 2   THE COPPER CYCLE

<u>Discussion</u>
Each reaction in this experiment goes to nearly 100% completion.   The steps and corresponding balanced chemical reactions are as follows:

1. The copper metal is reacted with concentrated nitric acid, HNO_3, to give a bright blue solution of copper (II) nitrate, $Cu(NO_3)_2$.

$$Cu_{(s)} + 4\,HNO_{3(aq)} \rightarrow Cu(NO_3)_{2(aq)} + 2\,NO_{2(g)} + 2\,H_2O_{(l)}$$

2. The solution of $Cu(NO_3)_2$ is reacted with NaOH to produce the royal blue solid, copper (II) hydroxide, $Cu(OH)_2$.

$$Cu(NO_3)_{2(aq)} + 2\,NaOH_{(aq)} \rightarrow Cu(OH)_{2(s)} + 2\,NaNO_{3(aq)}$$

3. Upon heating, the $Cu(OH)_2$ is converted to black copper (II) oxide, CuO.

$$Cu(OH)_{2(s)} \rightarrow CuO_{(s)} + H_2O_{(l)}$$

4. The CuO reacts with hydrochloric acid, HCl, giving a pale blue-green solution of copper (II) chloride, $CuCl_2$.

$$CuO_{(s)} + 2\,HCl_{(aq)} \rightarrow CuCl_{2(aq)} + H_2O_{(l)}$$

5. The Cu^{2+} ions in the aqueous $CuCl_2$ solution are reduced to Cu metal by the addition of zinc metal.

$$CuCl_{2(aq)} + Zn_{(s)} \rightarrow ZnCl_{2(aq)} + Cu_{(s)}$$

6. Excess Zn metal is removed by reaction with HCl (from step 4, and added HCl) to give $ZnCl_2$ and H_2.

$$Zn_{(s)} + 2\,HCl_{(aq)} \rightarrow ZnCl_{2(aq)} + H_{2(g)}$$

<u>Chemicals</u>
    Copper metal, $Cu_{(s)}$
    Zinc metal, $Zn_{(s)}$
    Sodium hydroxide, $NaOH_{(aq)}$, 6 M
    Nitric acid, $HNO_{3(aq)}$, 8 M (very corrosive: handle with care)
    Hydrochloric acid, $HCl_{(aq)}$, 6 M
    Acetone
    $NO_{2(g)}$ is a byproduct of the reaction

<u>Waste</u>
    All liquid waste can be disposed down the drain; dispose of the copper in the trash.

24

Procedure

1. Assemble the trap apparatus shown in Figure 2.1. The gas trap should contain about 150 mL of tap water <u>plus about 10 mL of 6 M NaOH solution</u>. **CAUTION:** When the Cu reacts with HNO_3, the noxious gas NO_2 is produced. The apparatus traps the gas and the alkaline sodium hydroxide reacts quickly with the NO_2, an acidic substance, keeping it from escaping into the laboratory atmosphere.

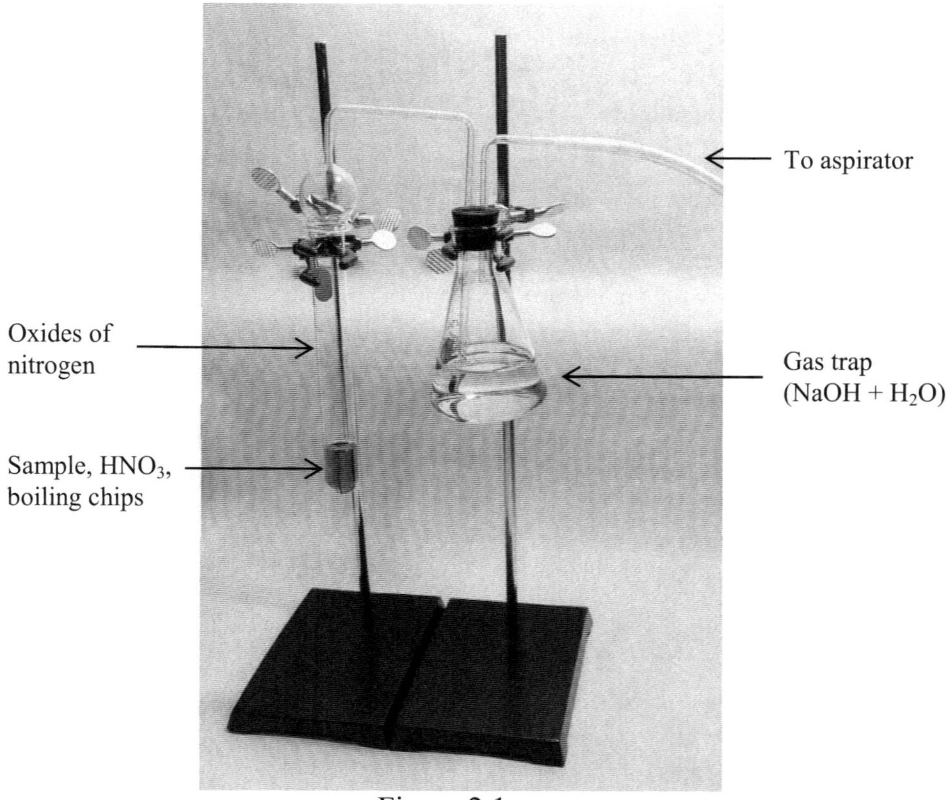

Figure 2.1
Gas Trap Set Up

2. Connect the trap outlet to the aspirator (sidearm of the water faucet), and turn on the water. Bubbles should be seen in the trap solution. Continue to apply suction to the system until all NO_2 has been reacted (through step 4).

3. Weigh the copper wire and record this mass. Place it in the 70 mL test tube. Add one or two boiling chips to ensure smooth boiling. Secure the test tube with the clamp at the <u>very top of the test tube</u>.

4. Add 10 mL (about an inch in a 70 mL test tube) of 8 M HNO_3. Relocate the glass hood over test tube before heating. Gently heat the sample using a Bunsen burner. Take away the flame from the sample if the bubbles above the solution are greater than 1 inch. Completely dissolve the sample. Continue heating until all the brown NO_2 gas on the sides of the test tube and greenish tint of the $Cu(NO_3)_2$ solution is gone. The resulting solution of $Cu(NO_3)_2$ should be a bright blue color. Let the solution cool.

$$Cu_{(s)} + 4\,HNO_{3(aq)} \rightarrow Cu(NO_3)_{2(aq)} + 2\,NO_{2(g)} + 2\,H_2O_{(l)}$$

5. Transfer the solution to a clean 250 mL beaker. Do not transfer the boiling chips (use a stirring rod to hold them back). Rinse the boiling chips and stirring rod with a small amount of DI water and add the water to the beaker. Dilute the total volume to about 50 mL with DI water.

6. Add about 20 mL of 6 M NaOH solution in 5 mL increments, slowly and with stirring (use the glass end of the stirring rod). Ensure the solution is very basic, as indicated by a deep blue color when tested with pH paper. Do not add more than 20 mL of NaOH. All of the Cu^{2+} should precipitate as $Cu(OH)_2$, a bright blue precipitate. Refer to the Appendix on how to test for pH using pH paper.

$$Cu(NO_3)_{2(aq)} + 2\,NaOH_{(aq)} \rightarrow Cu(OH)_{2(s)} + 2\,NaNO_{3(aq)}$$

7. Add about 80 mL of deionized water and boil gently using a hot plate, with continuous stirring to prevent bumping, until the blue $Cu(OH)_2$ is converted to the black CuO. When the solution is completely black, shut off the hot plate and remove the beaker to the lab bench using a hot mitt, and allow the solution to settle. As the precipitate sits in the hot aqueous solution, the growth of large particles of CuO at the expense of smaller ones takes place. The particle size will increase which will make the precipitate easier to filter. In this process, known as "digestion", small particles dissolve faster than larger ones, and are reprecipitated as larger particles.

$$Cu(OH)_{2(s)} \rightarrow CuO_{(s)} + H_2O_{(l)}$$

8. Discard the contents of the gas trap (in the 250 mL Erlenmeyer flask) to the drain in preparation to receive the filtrate. Prepare a funnel for gravity filtration by folding a large filter paper into a cone and placing it in the funnel. Moisten it with DI water to seal it to the funnel. Let the funnel rest in the empty 250 mL Erlenmeyer flask.

9. After the CuO has settled and the beaker is cool enough to handle without mitts, quantitatively transfer the precipitate and supernatant (liquid above the precipitate) through the funnel to the Erlenmeyer flask. Use a wash bottle in transferring the precipitate to wash down the sides of the beaker, but it is not necessary to remove all of the CuO from the beaker at this step. <u>Discard the colorless filtrate to the drain</u>.

10. When almost all the liquid is gone from the funnel, pour about 30 mL of 6 M HCl through the filter, catching the solution in the Erlenmeyer flask. Pour the same solution quantitatively back through the filter and catch it in the 250 mL reaction beaker. Repeat this process, pouring the solution back and forth through the filter, until all of the solid CuO has dissolved (from the beaker sides and from the filter paper) and all the solution is in the beaker. Finally, rinse the filter paper with a small amount of deionized water, adding the rinsing liquid to the solution in the beaker.

$$CuO_{(s)} + 2\,HCl_{(aq)} \rightarrow CuCl_{2(aq)} + H_2O_{(l)}$$

11. Place a <u>small</u> scoop of Zn powder on a piece of weighing paper. Add the Zn metal to the copper solution, a little at a time, with continuous stirring, until the solution has no blue or green tint (essentially colorless) and the excess Zn has reacted (this is

indicated by the cessation of H_2 bubbles). Add more HCl if necessary to speed up the removal of the excess zinc. Allow the copper to settle.

$$CuCl_{2(aq)} \; + \; Zn_{(s)} \; \rightarrow \; ZnCl_{2(aq)} \; + \; Cu_{(s)}$$

12. Set up the suction filtration apparatus shown in Figure 2.2. Mark a piece of filter paper that fits the Büchner funnel with pencil (not pen) to identify it, and then weigh a watch glass with the filter paper. Place the filter paper (marked side down) in the funnel. Turn on the vacuum (aspirator) and wet the filter paper down with deionized water.

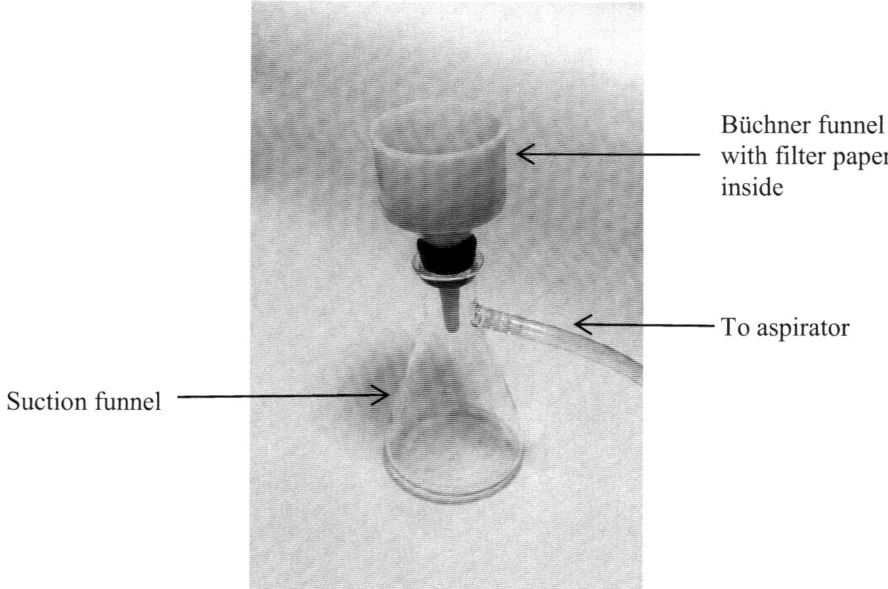

Figure 2.2
Büchner Funnel Set Up

13. Quantitatively transfer the copper in the beaker to the Büchner funnel. To do this, decant most of the liquid through the funnel and after most of the liquid is gone, begin to transfer the precipitate to the <u>center</u> of the filter paper. <u>Do not allow the liquid to fill the funnel as this will cause some of the copper to adhere to the sides of the funnel and not be captured on the filter paper.</u> Use small portions of water from a wash bottle and the rubber policeman to help in transferring the precipitate.

14. After the precipitate has been transferred, break up any large clumps of copper with a stirring rod, and then wash it several times with acetone. Continue to pull air through the funnel until the odor of acetone can no longer be detected. **<u>CAUTION</u>**: Acetone is flammable.

15. Turn off the suction to the apparatus. Transfer the copper and the filter paper to the watch glass by placing the watch glass over the Büchner funnel and then inverting the Büchner funnel. If the precipitate and paper are sufficiently dry they should easily fall out of the funnel onto the watch glass. Discard the filtrate to the drain.

16. Break up any large clumps of copper. Place the watch glass containing the filter paper and precipitate in an oven at >85°C for about 10 min. Remove, cool to room

temperature, and weigh. Return the watch glass with the filter paper and copper to the oven and reheat for about 5 minutes. Remove it again, cool at least 5 minutes, and reweigh. Repeat until a constant weight is attained ($\pm$ 0.01 g). Record the masses on the laboratory report sheet. Dispose of the precipitate in the trash.

17. From the original mass of the copper and the mass of the copper at the end of the cycle, calculate the percent recovery of copper.

$$\text{percent recovery} = \frac{\text{grams of copper recovered}}{\text{grams of initial sample}} \times 100$$

Many experiments in chemistry ask for the amount of a reactant used or product formed based on the known amount of another reactant or product in the equation.

Example

How many grams of barium phosphate are produced from 1.500 g of barium chloride? The balanced chemical reaction is as follows:

$$3\ BaCl_2\ {}_{(aq)} + 2\ Na_3PO_4\ {}_{(aq)} \rightarrow Ba_3(PO_4)_2\ {}_{(s)} + 6\ NaCl\ {}_{(aq)}$$

Strategy:
1. Start with the compound for which two things are known. From this information, calculate the moles of that compound. (In this example, it is $BaCl_2$: its mass and molar mass are known.)

2. Use the mole ratios in the balanced chemical reaction to convert to any other species in the reaction. (The mole ratio in this example is 3 moles $BaCl_2$ to 1 mole $Ba_3(PO_4)_2$.)

3. Obtain the final answer using the molar mass of the "other" compound (or use molarity if a volume is needed, a concept that will be introduced in later labs.)

- Always use dimensional analysis as shown in this example.
- When using molar mass, the numerical value stays with the "g" unit.

$$1.500\ g\ BaCl_2 \times \frac{1\ \text{mole}\ BaCl_2}{208.20\ g\ BaCl_2} \times \frac{1\ \text{mole}\ Ba_3(PO_4)_2}{3\ \text{mole}\ BaCl_2} \times \frac{601.84\ g\ Ba_3(PO_4)_2}{1\ \text{mole}\ Ba_3(PO_4)_2} = 1.445\ g\ Ba_3(PO_4)_2$$

LABORATORY REPORT SHEET

Name____ Date________________

Partner_______________________________________

**READ ALL EQUIPMENT TO
THE CORRECT PLACE VALUE
AND RECORD ALL VALUES
WITH PROPER UNITS!**

1. Mass of initial Cu   _______________

2. Mass of watch glass + filter paper   _______________

3. Mass of watch glass + filter paper + Cu recovered

      1st weighing   _______________

      2nd weighing   _______________

      3rd weighing, if necessary   _______________

4. Mass of Cu recovered   _______________

5. Percent recovery   _______________

Error Analysis

1. List two sources of error that would make the percent recovery higher than 100%.

2. List two sources of error that would make the percent recovery less than 100%.

3. Write an equation from this experiment that is a double displacement reaction.

4. Write an equation from this experiment that is a single displacement reaction *and* an oxidation-reduction reaction.

5. Write an equation from this experiment that is a decomposition reaction.

6. In this experiment, copper (II) oxide is reacted with hydrochloric acid. Sulfuric acid can be used instead of hydrochloric acid. Write the balanced chemical equation, including states, for this reaction.

7. In this experiment, copper chloride is reduced by zinc. Referring to the activity series of metals (see Appendix), all metals "above" copper in the series will reduce the copper chloride to copper. Aluminum is one of these metals. Write the balanced chemical equation, including states, for this reaction.

8. In this experiment, the excess zinc is oxidized and made aqueous with the addition of hydrochloric acid. Other strong acids, such as cold sulfuric acid, may be used in this step, but warm nitric acid cannot be used. Why and what impact would this have on the percent recovery if warm nitric acid was used?

Name _______________________________

HOMEWORK EXERCISES

1. There are five chemical reactions in this experiment that involve a copper compound. For each reaction,
   - Write the reaction in words, using the correct nomenclature.
   - Write the balanced equation using chemical formulas, including states.
   - On the product side of the reaction, clearly indicate the color of the copper precipitate or copper solution under the compound giving the color.
   - Write the reactions in the chronological order they will be performed. The first reaction is given as an example.

<u>Reaction #1</u>:

Copper reacts with nitric acid to give copper (II) nitrate and nitrogen dioxide and water

$$Cu_{(s)} + 4\,HNO_{3(aq)} \rightarrow Cu(NO_3)_{2(aq)} + 2\,NO_{2(g)} + 2\,H_2O_{(l)}$$

Bright blue solution

<u>Reaction #2</u>:

<u>Reaction #3</u>:

<u>Reaction #4</u>:

<u>Reaction #5</u>:

Continued on back

2. The copper is dissolved to start this experiment.
    a. What reagent is used to dissolve the copper?________________

    b. What toxic gas is evolved in this reaction?________________

    c. What is the function of the sodium hydroxide in the gas trap?

    d. What indicates the successful removal of the toxic gas?

3. a. What reagent is used to reduce the Cu^{2+} to Cu in this experiment? ____________

    b. Write the balanced chemical equation for removing this excess reducing agent. Include states.

    c. What indicates that all the Cu^{2+} in solution has been converted to Cu?

4. a. Between what two values must the percent recovery of any experiment lie?

    b. Calculate the percent recovery of copper if 1.9645 g of copper are recovered from 2.3872 g of copper wire.

5. Use the reactions from Question 1 of this homework. <u>Use dimensional analysis</u>.
    a. How many grams of nitric acid are required to react with 2.75 g of copper?

    b. How many grams of toxic gas are formed from 2.75 g of copper?

    c. What is the minimum amount of zinc required to precipitate 1.4621 g of copper?

EXPERIMENT 3
STOICHIOMETRIC CALCULATIONS AND GRAPHING

Ethanol solutions of different concentration are reacted with potassium dichromate. The absorbance of each solution is measured, and a calibration graph is constructed. The absorbance of an unknown ethanol solution is tested, and the concentration of ethanol is determined using the calibration graph.

Key Chemical Reactions:

$$2\ K_2Cr_2O_{7\ (aq)} + 8\ H_2SO_{4\ (aq)} + 3\ C_2H_5OH_{\ (aq)} \rightarrow$$
$$2\ Cr_2(SO_4)_{3\ (aq)} + 2\ K_2SO_{4\ (aq)} + 3\ CH_3COOH_{\ (aq)} + 11\ H_2O_{\ (l)}$$

Key Mathematical Equations:

dilution formula: $\quad C_iV_i = C_fV_f$

dimensional analysis:

$$0.0250\ \text{g}\ C_2H_5OH\ \times\ \frac{1\ \text{mole}\ C_2H_5OH}{46.08\ \text{g}\ C_2H_5OH}\ \times\ \frac{2\ \text{moles}\ K_2Cr_2O_7}{3\ \text{mole}\ C_2H_5OH}\ \times\ \frac{294.20\ \text{g}\ K_2Cr_2O_7}{1\ \text{mole}\ K_2Cr_2O_7} = 0.106\ \text{g}\ K_2Cr_2O_7$$

<u>Discussion</u>

The fundamental chemistry in this procedure is the basis of the Breathalyzer test, a test used to determine the level of alcohol intoxication of a person. The test is shown by the following chemical reaction:

$$2\ K_2Cr_2O_{7\ (aq)} + 8\ H_2SO_{4\ (aq)} + 3\ C_2H_5OH_{\ (aq)} \rightarrow$$
$$2\ Cr_2(SO_4)_{3\ (aq)} + 2\ K_2SO_{4\ (aq)} + 3\ CH_3COOH_{\ (aq)} + 11\ H_2O_{\ (l)}$$

The alcohol (ethanol, C_2H_5OH) reacts with the orange Cr^{6+} ion in $K_2Cr_2O_7$, producing the green Cr^{3+} ion of $Cr_2(SO_4)_3$. The solution is exposed to a light source in a spectrometer (I_0), and it absorbs some of this light depending upon the amount of Cr^{3+} present. The amount of light transmitted (I) through the sample at a specific wavelength is converted into an electrical signal. This is converted to an absorbance value, which is proportional to the amount of alcohol present in the sample. The schematic of a spectrometer is shown in Figure 3.1.

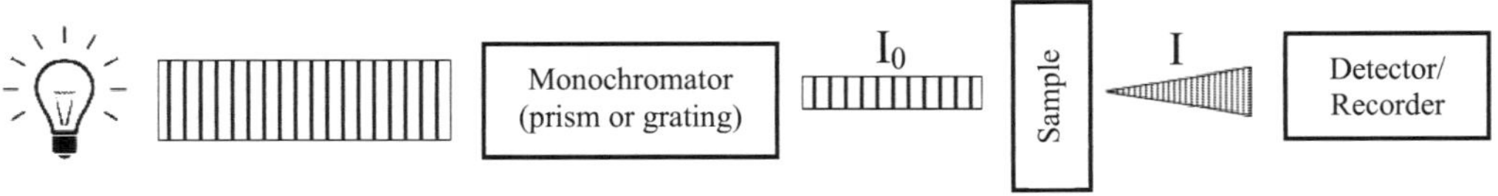

Figure 3.1. Schematic diagram of a spectrophotometer

EXPERIMENT 3   STOICHIOMETRIC CALCULATIONS AND GRAPHING

Instruments, such as a spectrometer, must be calibrated to assure the operator that the results obtained are accurate.   In this experiment, a calibration graph relating the absorbance of solutions to % ethanol will be constructed with known values to insure the instrument's reliability.   An unknown sample will then be prepared and the % ethanol will be determined using this calibration graph.

There are three different skills that will be developed in this lab: graphing a set of data, use of the dilution formula, and practice with stoichiometric calculations.

Graphing
In chemistry, data is collected, and often a graph, or a visual representation of the data, can be constructed showing the relationship between the data points. To properly graph, use the following procedure:

Identify the Independent and Dependent Variables
The independent variable is the variable that is intentionally manipulated in the experiment. These values are represented on the x-axis (horizontal axis, or abscissa). The dependent variable is the data collected as the independent variable is deliberately changed. These values are represented on the y-axis (vertical axis, or ordinate).

Appropriately Scale the Axes
The area of the plot should take up most of the graph paper. If restricted to a smaller part of the paper, much accuracy in the determination of specific values on the curve will be lost. This is shown below in Figures 3.2A, 3.2B, and 3.2C.

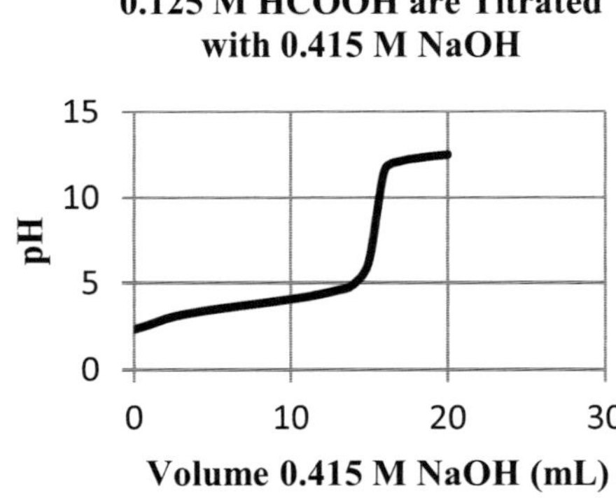

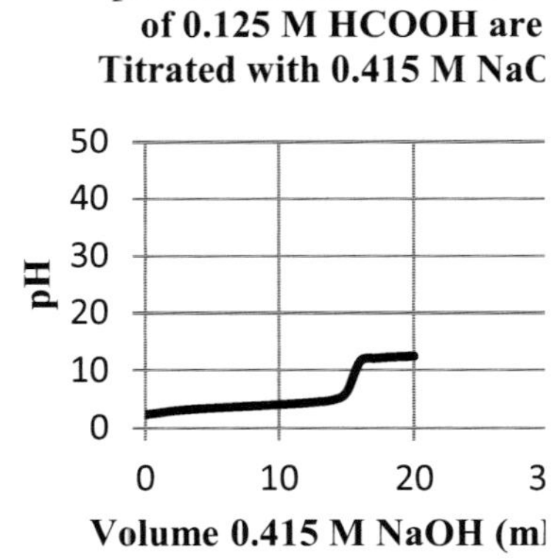

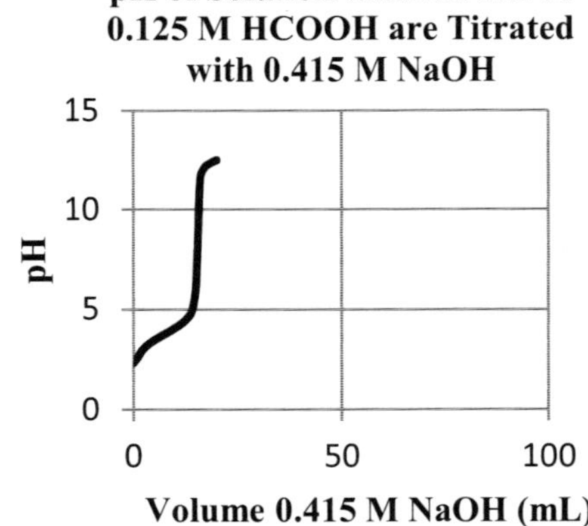

Figure 3.2A. Good scaling on both axes

Figure 3.2B. Poor scaling on y-axis

Figure 3.2C. Poor scaling on x-axis

To select a scale for each axis, take the range of the data and divide by the number of squares available. Choose major divisions that make smaller divisions easy to determine and the data easy to interpret. The origin of each axis does not always have to start at zero.

For example, assume 210 mL of NaOH were delivered in a titration experiment, and the mL of NaOH was determined to be the independent variable. The graph paper has 10 major divisions on the x axis with 10 smaller divisions within each major one. (This is the scale of the graph paper found at the end of this manual.) To scale, divide 210 by 10 (major divisions), obtaining 21. Round this UP to a convenient number, such as 25. Each major division is now 25 mL, and each minor division is 2.5 mL. <u>At least 70% of the axis should be utilized</u>.

Labeling Axes and Title
<u>Axes</u>
- Label each axis as descriptively as possible, including units, where applicable.
- If the unit is a common unit, such as K, mL, g/mol, etc., put the units in parenthesis.

<u>Title</u>
- Label the title descriptively, including as much information about the two variables as possible.
- The title should always be in the form "y vs x", i.e., dependent variable label vs. independent variable label.
- Spell out any abbreviations that may have been used in the axes labels.
- It is not necessary to include common units from the axes labels.

Two examples are given in Figure 3.3.

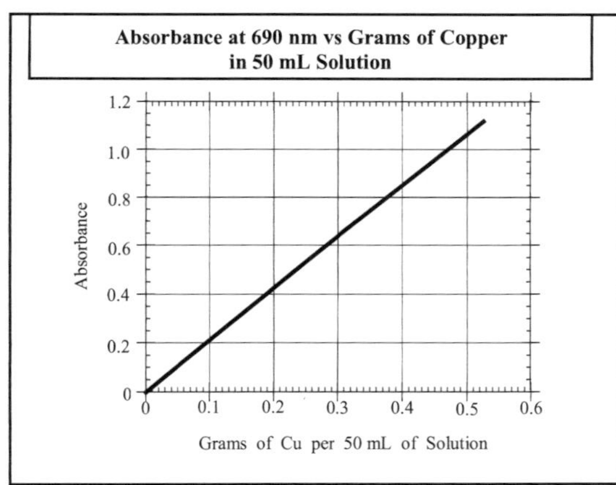

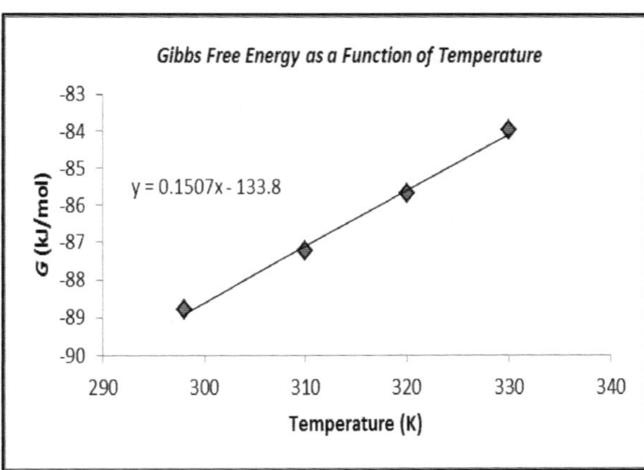

Figure 3.3: Examples of proper labelling of a graph's axes and title.

Drawing Curves or Lines
Draw a best fit curve for the data, making the line as smooth as possible. If the data is supposed to be linear, draw a line that passes near most of the points. (It does not have to go through the centers of the points, and some points may be invalid.) If the data is supposed to be curved, use a French curve to draw a smooth line through the points. Avoid simply connecting the dots.

Extrapolation

Sometimes a value is desired that is not contained in the part of the graph depicted. For linear graphs (only), extrapolation can be used. To do this, first find the slope of the line, and write the equation of the line. Substitute in the x or y point not shown on the graph and calculate the other variable.

Example: In the right graph of Figure 3.3, determine the value of G at 400 K.

The slope of the line is 0.1507, and the equation of the line is $y = 0.1507x - 133.8$

The x-value is the temperature, and the y-value is G. Substituting in 400 K:

$$y = 0.1507 (400) - 133.8 = -75.5 \text{ kJ/mol}$$

The Dilution Formula

Dilution is a procedure that reduces the concentration of a solution. It is prepared using a concentrated solution (or stock solution) and then adding a solvent, lowering the concentration in the final solution. The final concentration of the dilute solution can be calculated using the dilution formula:

$$C_i V_i = C_f V_f$$

Where
C_i = the initial concentration of the stock solution C_f = the final concentration of the diluted solution
V_i = the volume of the stock solution used V_f = the total volume of the diluted solution

Concentration can be expressed in various ways, including mass, %, moles, molarity, etc. It does not matter what units of concentration, mass, or volume are used, as long as they are consistent on both sides of the equation. In this experiment, % by volume will be used.

The dilution equation can be used ONLY when reducing the concentration of a solution containing the SAME component. It CANNOT be used in stoichiometric calculations, and will be counted wrong if used.

Example

10.0 mL of a 25.0% by mass sulfuric acid solution are diluted with water to a volume of 50.0 mL. What is the concentration of the new solution?

$$C_i V_i = C_f V_f$$

$$(25.0\%)(10.0 \text{ mL}) = (C_f)(50.0 \text{ mL})$$

$$C_f = 5.00\%$$

EXPERIMENT 3   STOICHIOMETRIC CALCULATIONS AND GRAPHING

<u>Stoichiometric Calculations</u>

Many experiments in chemistry ask for the amount of a reactant used or product formed based on the known amount of another reactant or product in the equation. Use the following strategy:

1. Start with the compound for which two things are known to get that compound into moles. (Perhaps its mass and molar mass or its volume and density are known.)
2. Use the mole ratios in the balanced chemical reaction to convert to any other species in the reaction.
3. Obtain the final answer of the "other" compound
   - Use the molar mass to obtain grams.
   - Use the molar mass, then density if a volume is needed. Note: Density can only be used for pure liquids, solids, and gases. It cannot be used for aqueous solutions. For aqueous solutions, molarity is used.
   - Always use dimensional analysis, clearly showing what is in the numerator and denominator, so it is obvious that units cancel out.
   - The numerical value stays with the numerator unit.
   - Include ALL units and labels (unit and compound formula).

Example

How many grams of iodine are formed from 5.4167 g of potassium permanganate? How many mL of water are formed? The density of water is 1.00 g/mL. The balanced chemical reaction is as follows:

$$10\ HI\ _{(aq)} + 2\ KMnO_4\ _{(aq)} + 3\ H_2SO_4\ _{(aq)} \rightarrow 5\ I_2\ _{(aq)} + 2\ MnSO_4\ _{(s)} + K_2SO_4\ _{(aq)} + 8\ H_2O\ _{(l)}$$

$$5.4167\ g\ KMnO_4 \times \frac{1\ mole\ KMnO_4}{158.04\ g\ KMnO_4} \times \frac{5\ mole\ I_2}{2\ mole\ KMnO_4} \times \frac{253.80\ g\ I_2}{1\ mole\ I_2} = 21.747\ g\ I_2$$

$$5.4167\ g\ KMnO_4 \times \frac{1\ mole\ KMnO_4}{158.04\ g\ KMnO_4} \times \frac{8\ mole\ H_2O}{2\ mole\ KMnO_4} \times \frac{18.02\ g\ H_2O}{1\ mole\ H_2O} \times \frac{1.00\ mL}{1.00\ g} = 2.47\ mL\ H_2O$$

<u>Chemicals</u>

Ethanol solution	C_2H_5OH
Dichromate solution	$K_2Cr_2O_7, AgNO_3, H_2SO_4$

(contains potassium dichromate, silver nitrate and sulfuric acid)

<u>Waste</u>

All waste is to be collected and disposed in the hazardous waste drum. Anything that contains the dichromate solution CANNOT be disposed to the drain.

<u>Procedure</u>

<u>Calibration Curve</u>

1. Obtain 0.125 M $K_2Cr_2O_7$ solution in a small beaker and normalize a 10.00 mL pipet with the solution. Dispose the normalizing liquid to waste. Pipet 10.00 mL of the dichromate solution to a clean 125 mL Erlenmeyer flask.

2. Fill a small (10-30 mL) beaker with DI water.

3. <u>As assigned by the instructor</u>, make ONE of the following solutions in the 125 mL Erlenmeyer flask containing the dichromate solution. Use clean disposable pipets to dispense drops of each component.

Solution #	1	2	3	4
drops 10.0 % ethanol solution	5	10	15	20
drops water	15	10	5	0

4. Allow the chemicals to react for five minutes, swirling the flask occasionally.

5. Add 39.0 mL of DI water to the Erlenmeyer flask. Swirl the flask gently. The volume in the flask is now 50.0 mL.

6. Bring the volumetric flask to the spectrometer, where the instructor or TA can help with the following. The spectrometer has already been calibrated.
   a. Normalize the cuvet (small plastic rectangular prism) with your solution. Dispose of the solution to waste.
   b. Fill the cuvet to ½" from the top. Wipe the sides with a Kim-wipe to remove fingerprints and any drops of liquid. Make sure there are no bubbles in the cuvet (if there are, gently tap the cuvet on the lab bench).
   c. Record the absorbance of the solution on your data sheet. Also record the number on the board for the whole class.
   d. Dispose of the contents of the cuvet to waste.

7. Construct a graph of the data of the following two parameters: absorbance and % ethanol in final prepared solution. Plot each absorbance value obtained by all groups in the class.

<u>Determination of volume % Ethanol in an Unknown</u>
1. Pipet 10.00 mL of the 0.125 M $K_2Cr_2O_7$ solution to a clean 125 mL Erlenmeyer flask.

2. Add 20 drops of the unknown to the flask.

3. Allow the chemicals to react for five minutes, swirling occasionally.

4. Add 39.0 mL of DI water to the Erlenmeyer flask. Swirl the flask gently. The volume in the flask is now 50.0 mL.

5. Bring the volumetric flask to the spectrometer, and determine the absorbance of the solution, following step 6 above. Dispose of the contents of the cuvet to waste. Using the calibration graph, determine the % ethanol in the unknown.

LABORATORY REPORT SHEET

Name____   Date__________________   <u>Note: 20 drops = 1.0 mL</u>

Solution #		1	2	3	4
drops 10. % ethanol solution	0	5	10	15	20
drops water	20	15	10	5	0
% ethanol in 1 mL (20 drops)	(10. %)(0 drops) = (y)(20 drops) ethanol total y = 0.0%				
% ethanol in final prepared solution Use the dilution formula and show work, with units	(0.0%)(1.0 mL) = (x)(50.0 mL) initial V final V x = 0.0%				
Absorbance values at 580 nm *Record each group's values as shown on the board*					
Color of solution					

Unknown Number   _____________

Absorbance   _____________

% ethanol in unknown   _____________________

Conclusion: In a short paragraph written in a technical manner, summarize in two or three sentences the purpose of the lab. Next, in sentence form, detail <u>relevant</u> data from the laboratory report sheet, in logical order, including units where appropriate. Also include any color changes that happened during the reaction.

Calculations

The reaction taking place in the experiment is as follows. Write the name of the reagent or product in the provided space below each component.

$$2\ K_2Cr_2O_{7\ (aq)} +\ 8\ H_2SO_{4\ (aq)} +\ 3\ C_2H_5OH_{\ (aq)} \rightarrow\ 2\ Cr_2(SO_4)_{3\ (aq)} +\ 2\ K_2SO_{4\ (aq)} + 3\ CH_3COOH_{\ (aq)} +\ 11\ H_2O_{\ (l)}$$

Molar mass (g/mol)						
294.20	98.09	46.08	392.21	174.27	60.06	18.02

					acetic acid	

Record the % ethanol in your final prepared solution (NOT the unknown). This was assigned by the instructor.

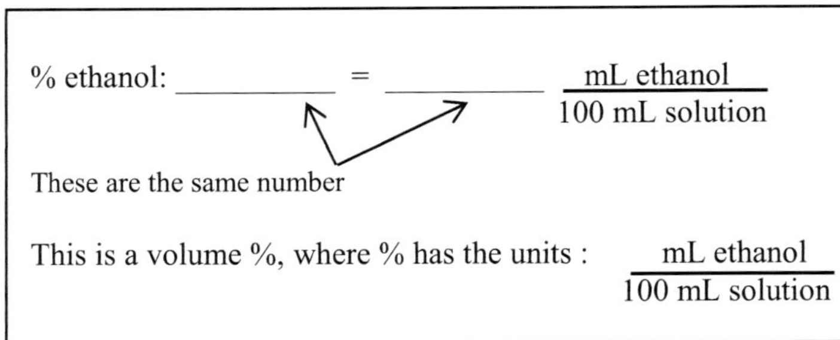

1. Based on the value in the box, complete the equation below to calculate the grams of ethanol in your final prepared solution (50.0 mL). The density of ethanol is 0.789 g/mL.

$$\frac{\_\_\_\_\_\_\ mL\ ethanol}{100\ mL\ solution} \times 50.0\ mL\ solution \times \frac{0.789\ g\ ethanol}{mL} = \_\_\_\_\_\_\ g\ ethanol$$

Based on the value (grams of ethanol) in Calculation 1 above, calculate the following:

2. Calculate the grams of $K_2Cr_2O_7$ reacted.

3. Calculate the grams of H_2SO_4 reacted.

4. Calculate the grams of $Cr_2(SO_4)_3$ formed.

5. Calculate the grams of K_2SO_4 formed.

6. Calculate the grams of acetic acid formed.

7. Calculate the volume in mL of water formed. The density of water is 1.00 g/cm^3.

Questions and Error Analysis:
1. Circle any points that you feel are outliers on your graph. Should outliers be considered in drawing a line of best fit? Why or why not?

2. Why is a calibration curve necessary?

3. In this experiment, is a visual knowledge of the color of the solution necessary? Why or why not?

4. If the solution was diluted in the 125 mL Erlenmeyer flask to a volume <u>more</u> than 50.0 mL how would this affect absorbance and why?

5. a. Determine the equation for the line drawn on your graph.

    b. Extrapolate and find the % ethanol at an absorbance of 1.2

6. If the cuvet was not normalized, and the last solution in it was of a higher ethanol concentration, how would this affect the absorbance? Explain.

7. In a Breathalyzer test, two tests are usually done, and they must agree with one another. Why is this done?

This homework is four pages

Name _______________________

HOMEWORK EXERCISES

Dilution Formula

Calculate the concentration or volume in the diluted or concentrated solution. If the dilution formula cannot be used, write "Cannot be used" and state the reason.

1. Calculate the final concentration of the solution when 10.0 mL of an 18 M sulfuric acid solution are diluted with water to a total volume of 1.00 L.

2. 90.0 mL of water are added to 10.0 mL of a 5.5% ethanol solution. Calculate the concentration of the final solution.

3. 20.0 mL of a 1.0 M solution of acetic acid are titrated with 25.0 mL of NaOH. Calculate the concentration of the NaOH solution.

4. How much 50.0% NaOH is required to make 100.0 mL of 10.0% NaOH?

5. 25.0 mL of a concentrated solution of acetic acid was diluted to a volume of 50.0 mL. It was then titrated with 2.0 M NaOH and the concentration of the diluted acid was determined to be 0.347 M. What is the original concentration of the acid?

Stoichiometric Calculations   *Dimensional Analysis must be used to receive credit*

1. Balance the following equation and calculate the following:

$$\underline{\quad} \; NaN_3 \; \rightarrow \; \underline{\quad} \; Na \; + \; \underline{\quad} \; N_2$$

a. How many grams of nitrogen are formed from 15.65 grams NaN_3?

b. The density of nitrogen is 1.153×10^{-3} g/mL. How many m^3 of nitrogen are formed?

2. Balance the following equation and calculate the following:

$$\underline{\quad} \ CaCl_2 \ + \ \underline{\quad} \ Na_3PO_4 \ \rightarrow \ \underline{\quad} \ NaCl \ + \ \underline{\quad} \ Ca_3(PO_4)_2$$

a. How many grams of $CaCl_2$ are required to make 2.85 g $Ca_3(PO_4)_2$?

b. How many grams of Na_3PO_4 react with 4.1856 g $CaCl_2$?

c. How many grams of NaCl are formed from 3.6825 g Na_3PO_4?

3. Elemental aluminum reacts with iron (III) oxide forming aluminum oxide and elemental iron.
    a. Write the balanced chemical equation for this reaction.

b. How many grams of iron are formed from 5.1629 g of aluminum?

c. How many grams of aluminum oxide are formed from 5.1629 g of aluminum?

d. How many grams of iron (III) oxide react with 5.1629 g of aluminum?

EXPERIMENT 3   STOICHIOMETRIC CALCULATIONS AND GRAPHING

<u>Graphing, Extrapolation, and Scaling</u>

1. Label both axes (with units) and give each graph an appropriate title. Data is given in the middle of the graph.

<table>
<tr>
<td>Pressure in atmospheres is changed and the volume is measured in mL</td>
<td>pH is measured as mL of NaOH are added</td>
</tr>
<tr>
<td>The temperature in Kelvin is recorded over a period of time (seconds)</td>
<td>The density (g/mL) is measured on a set of solutions of known mass % water</td>
</tr>
<tr>
<td>The ln of the rate constant, k (no units), is recorded as a function of 1/T (T is in Kelvin)</td>
<td>As temperature (in Kelvin) changes, the Gibb's Free Energy (ΔG, kJ/mol) is calculated</td>
</tr>
</table>

2. Extrapolate to determine how many grams of copper are present (in 50 mL of solution) if the absorbance is 1.75. Show work:

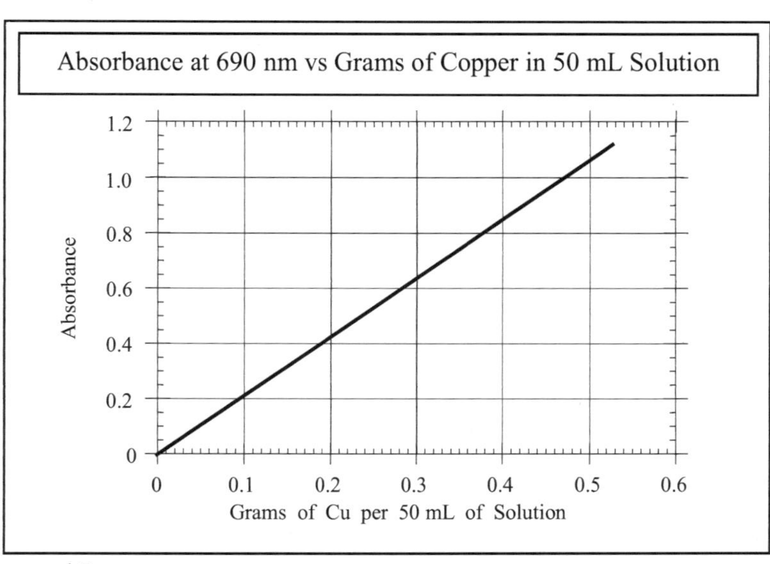

3. Scaling: There are 10 major divisions and within each major division, there are 10 minor divisions. Label the major divisions for the range indicated in each figure. The smaller divisions should have easily determinable/countable increments. Also record the major and minor increments.

At least 70% of the scale must be used by the range	Data Range	Major Division Increment	Minor Division Increment
Example: 0 150 300 450 600 750 900 1050 1200 1350 1500	Time 0-1500 seconds	150	15
	Volume 0-32 mL		
	Mass % 0-100%		
	Volume 0-21 mL		
	Density 0.80-1.25 g/mL		

4. Scaling: There are 7 major divisions and within each major division, there are 10 minor divisions. Label the major divisions for the range indicated in each figure. The smaller divisions should have easily determinable/countable increments. Also record the major and minor increments.

At least 70% of the scale must be used by the range	Data Range	Major Division Increment	Minor Division Increment
	pH 2-13		
	Density 0.78-1.05 g/mL		
	Temperature 8-27 °C		
	Temperature 280-350 K		
	Absorbance 0.01-1.15		

EXPERIMENT 4
LIMITING REAGENTS

Copper (II) chloride and aluminum are reacted together and the limiting reagent will be determined. The percent yield will be calculated and the amount of the unreacted excess reagent will also be determined.

Key Chemical Reactions:

$$3\ CuCl_2\ _{(aq)} + 2\ Al\ _{(s)} \rightarrow 2\ AlCl_3\ _{(aq)} + 3\ Cu\ _{(s)}$$

Key Mathematical Equations:

$$\% \text{ yield} = \frac{\text{actual yield}}{\text{theoretical yield}} \times 100$$

Dimensional analysis:

$$1.50 \text{ g } CuCl_2 \cdot 2\ H_2O \times \frac{1 \text{ mole } CuCl_2 \cdot 2\ H_2O}{170.48 \text{ g } CuCl_2 \cdot 2\ H_2O} \times \frac{1 \text{ mole } CuCl_2}{1 \text{ mole } CuCl_2 \cdot 2\ H_2O} \times \frac{3 \text{ mole } Cu}{3 \text{ mole } CuCl_2} \times \frac{63.55 \text{ g } Cu}{1 \text{ mole } Cu} = 0.559 \text{ g } Cu$$

Foundational Chemistry Concept: Solubility Rules for Common Ionic Compounds in Water
Mastery of Table 4.1 is required

<u>Discussion</u>
In a laboratory or industrial chemical reaction, the reactants are not added in their precise stoichiometric ratios. Usually one reactant is consumed before the other. This reactant is known as the <u>limiting reagent</u>. The other reactant(s) present are in excess, and some quantity of them is left over after the reaction is complete. These are called <u>excess reagents</u>. The amounts of the reactants are determined so that the yield of the desired product is maximized, while the cost is kept to a minimum. The limiting reagent is selected based on cost, availability, or reaction stoichiometry.

This experiment contains other terms that are used both in the laboratory and industry in addition to the terms <u>limiting reagent</u> and <u>excess reagent</u>. A <u>reactant</u> is a component on the left side of the balanced chemical equation and is a starting material of the process. A <u>product</u> is a component on the right side of the balanced chemical equation, and is a compound or element produced in a chemical reaction. An insoluble reaction product is called a <u>precipitate</u>. When separating a precipitate from a solution in a laboratory, the solution is called the <u>supernatant</u>, and it is <u>decanted</u>, or poured off.

The <u>% yield</u> of a reaction is defined as follows:

$$\% \text{ yield} = \frac{\text{actual yield}}{\text{theoretical yield}} \times 100$$

47

where the <u>actual yield</u> is the amount of the product experimentally collected and the <u>theoretical yield</u> is the amount of the product that would be obtained if every molecule of the limiting reagent was reacted and all the product collected. The % yield of a reaction can range from 0 to 100%. The <u>% purity</u> of a compound is the percentage of the pure compound present in an impure sample.

Example
0.8653 g of magnesium nitrate hexahydrate and 0.7215 g of potassium hydroxide react in aqueous solution.

   a. Predict the products, and write the balanced molecular equation (including states) for this reaction.
   b. Determine the limiting reactant and the amount of precipitate that should be formed (theoretical yield).
   c. Determine the amount of excess reactant present at the end of the reaction.
   d. Determine the % yield of the reaction if 0.1841 g of the precipitate are isolated.

a. Predict the products, and write the balanced molecular equation (including states) for this reaction.

To predict the products in aqueous solution, it is helpful to first break up the reactants into their ionic components by referring to a solubility chart, as shown in Table 4.1.

Table 4.1
Solubility Rules for Common Ionic Compounds in Water

Soluble	Insoluble
Group I and NH_4^+	Carbonates (CO_3^{2-})
Nitrates (NO_3^-)	Phosphates (PO_4^{3-})
Acetates (CH_3COO^-)	Chromates (CrO_4^{2-})
Bicarbonates (HCO_3^-)	Sulfides (S^{2-})
Chlorates and Perchlorates	
(ClO_3^- and ClO_4^-)	Except Group I and NH_4^+
Halides (Cl^-, Br^-, I^-)	
Except Ag^+, Hg_2^{2+}, Pb^{2+}	
Sulfates (SO_4^{2-})	Hydroxides (OH^-)
Except Ag^+, Ca^{2+}, Sr^{2+}, Ba^{2+}, Hg_2^{2+}, Pb^{2+}	Except Group I, NH_4^+, and Ba^{2+}

In a double displacement reaction like this, switch the positive and negative ions or polyatomic ions from each compound and then balance the new molecule, making sure the charges are balanced. In this example,

$Mg(NO_3)_2$ and KOH will dissociate in water to give the following ions

$Mg^{2+} + NO_3^-$ and $K^+ + OH^-$

EXPERIMENT 4   LIMITING REAGENTS

Switching the ions from one compound to the other gives

$$Mg^{2+} + OH^- \quad \text{and} \quad K^+ + NO_3^-$$

Balancing these using their charges gives the products

$$Mg(OH)_2 \quad \text{and} \quad KNO_3$$

The overall balanced equation then becomes

$$Mg(NO_3)_2 + 2\,KOH \rightarrow Mg(OH)_2 + 2\,KNO_3$$

The balanced molecular equation of the magnesium nitrate <u>hexahydrate</u> is then

$$Mg(NO_3)_2{\cdot}6\,H_2O_{(aq)} + 2\,KOH_{(aq)} \rightarrow Mg(OH)_{2\,(s)} + 2\,KNO_{3\,(aq)} + 6\,H_2O_{(l)}$$

Therefore, magnesium hydroxide is a precipitate (a reaction product that is not soluble), while the other compounds will all remain in solution.

b. *Determine the limiting reactant and the amount of precipitate that should be formed.*
To determine the limiting reagent, the amount of product (the magnesium hydroxide precipitate in this example) will be calculated using each reactant (two separate calculations). Whichever reactant makes the <u>least</u> amount of product will be the limiting reactant. Some quantity of the other reactant will be left over at the end of the reaction and is called the excess reagent.

$$0.8653\text{ g }Mg(NO_3)_2{\cdot}6\,H_2O \times \frac{1\text{ mole }Mg(NO_3)_2{\cdot}6\,H_2O}{256.3\text{ g }Mg(NO_3)_2{\cdot}6\,H_2O} \times \frac{1\text{ mole }Mg(OH)_2}{1\text{ mole }Mg(NO_3)_2{\cdot}6\,H_2O} \times \frac{58.3\text{ g }Mg(OH)_2}{1\text{ mole }Mg(OH)_2} = 0.1968\text{ g }Mg(OH)_2$$

$$0.7215\text{ g }KOH \times \frac{1\text{ mole }KOH}{56.1\text{ g }KOH} \times \frac{1\text{ mole }Mg(OH)_2}{2\text{ moles }KOH} \times \frac{58.3\text{ g }Mg(OH)_2}{1\text{ mole }Mg(OH)_2} = 0.3749\text{ g }Mg(OH)_2$$

Comparing these two results, the reactant that makes the least amount of product is $Mg(NO_3)_2 \cdot 6\,H_2O$. Therefore, it is the limiting reagent, and the maximum amount of $Mg(OH)_2$ that can be produced is 0.1968 g. Never assume that the reactant with the smaller initial mass is the limiting reagent!

Often, the reactants are in solution, and volumes (instead of masses) are known. To determine the limiting reagent in these examples, calculate moles of each reactant based on their volume and molarity, and continue through to the desired product.

c. *Determine the amount of excess reactant present at the end of the reaction.*
Determining the amount of excess reactant present at the end of the reaction is a two-step process. First, the amount of excess reactant consumed is calculated. When doing this, it is helpful to start with the initial amount of the limiting reagent and then calculate the amount of excess reagent needed to react with this. Then, this amount consumed is subtracted from the original excess reactant amount to obtain the amount left unreacted.

49

Amount of excess reactant consumed in the reaction:

$$0.8653 \text{ g Mg(NO}_3)_2 \bullet 6\text{ H}_2\text{O} \times \frac{1 \text{ mole Mg(NO}_3)_2 \bullet 6\text{ H}_2\text{O}}{256.3 \text{ g Mg(NO}_3)_2 \bullet 6\text{ H}_2\text{O}} \times \frac{2 \text{ moles KOH}}{1 \text{ mole Mg(NO}_3)_2 \bullet 6\text{ H}_2\text{O}} \times \frac{56.1 \text{ g KOH}}{1 \text{ mole KOH}} = 0.3788 \text{ g KOH}$$

Amount of excess reactant present after the reaction:

0.7215 g KOH (start) – 0.3788 g KOH (reacted) = 0.3427 g KOH (excess)

This can also be shown and summarized as shown below:

	$Mg(NO_3)_2$ (aq)	+	$2\,KOH$ (aq)	$\rightarrow$	$Mg(OH)_2$ (s)	+	$2\,KNO_3$ (aq)
	0.8653 g		0.7215 g		0 grams		0 grams
	↓		↓				
Initial	0.003376 mole		0.01286 mole		0 mole		0 mole
Change	– 0.003376 mole		– 0.006752 mole		+ 0.003376 mol		
Final	0 mole		0.006108 mole		0.003376 mole		
	↓		↓		↓		
	0 g		0.3427 g		0.1968 g		

d. *Determine the % yield of the reaction if 0.1841 g of the precipitate are isolated.*

$$\% \text{ yield} = \frac{\text{actual yield}}{\text{theoretical yield}} \times 100\,\% \qquad \frac{0.1841 \text{ g}}{0.1968 \text{ g}} \times 100 = 93.55\%$$

<u>Chemicals</u>
   Copper (II) chloride dihydrate $CuCl_2 \bullet 2\,H_2O$
   Aluminum foil   Al
   Hydrochloric acid, 6 M   HCl

<u>Waste</u>
   All aqueous waste can be disposed down the drain. Copper solids can be disposed in the trash.

<u>Procedure</u>
1. Weigh and accurately record approximately 0.5 g of $CuCl_2 \bullet 2\,H_2O$. Place in a 100 mL beaker.

2. Weigh and accurately record approximately 0.5 g of aluminum foil (~4" x 4").

3. Weigh an evaporating dish.

4. Add approximately 30 mL of DI water to the $CuCl_2 \cdot 2\ H_2O$. Swirl to dissolve the solid. It may be necessary to use a stirring rod to break up the solid. Use the DI water bottle to rinse off the stirring rod to the beaker. Note the color of the solution.

5. Add the aluminum foil to the beaker, roughly torn in large pieces and crumpled in loose balls.

6. After 3 minutes, slowly add approximately 20 mL of 6 M HCl.

7. Wait until all the bubbling has subsided. Gently break up any floating copper with a stirring rod to get all of the copper to settle. Do not pulverize it. Record the color of the solution.

8. When all the copper has settled, decant about half of the supernatant to a 400 mL beaker using this technique: put the stirring rod at a 90° angle in the lip of the 100 mL beaker, with the stirring rod extending into the 400 mL beaker, and pour the liquid out.

9. Add about 30 mL of water from the DI water bottle to the contents in the 100 mL beaker by squirting water on the side of the beaker (not directly into the solution). Let the copper settle again and <u>decant almost all of the supernatant</u>.

10. Repeat step 9 at least five more times (frequently up to ten times), until the water with the copper has a pH above 5.

11. Decant most of the water and transfer the copper to an evaporating dish. Use a small amount of water from the DI water bottle, squirting water on the sides of the beaker to completely transfer the copper. Use the stirring rod to break up large chunks of copper.

12. Place the evaporating dish on a ring clamp covered with wire gauze and heat the copper to dryness using a Bunsen burner. Heat for 1-2 minutes after the water has been evaporated. If the copper starts turning dark, remove the heat source and continue with the next step.

13. Shut off the Bunsen burner and remove the evaporating dish from the wire gauze with the hot hands to the laboratory bench.

14. When the evaporating dish is at room temperature, weigh the dish and copper.

LABORATORY REPORT SHEET

Name_______________________________________Date__________________

Partner _________________________________

Mass of copper (II) chloride dihydrate	
Mass of aluminum	
Mass of empty evaporating dish	
Mass of evaporating dish plus copper	
Mass of copper	

1. a.   Write the balanced molecular reaction for the reaction between copper (II) chloride dihydrate and aluminum in aqueous solution. Include states.

   b.   Write the balanced ionic and net ionic reactions (with states) for this reaction.

Ionic:

Net Ionic:

2.   Calculate the theoretical mass of copper produced based on the mass of the copper (II) chloride dihydrate.

3.   Calculate the theoretical mass of copper produced based on the mass of the aluminum.

4.   What is the limiting reagent? _______________________________

5.   How many grams of excess reagent are left over?

6.   What is the % yield of copper?

<u>Conclusion</u>:
In a short paragraph written in a technical manner, summarize in one or two sentences the general lab procedure. Next, in sentence form, detail <u>relevant</u> data from the laboratory report sheet, in logical order, including units where appropriate. Also include physical observations that happened during the reaction.

<u>Questions and Error Analysis</u>:

1.   What physical parameter confirms the reaction is complete and the supernatant can start to be decanted?

2.   What physical evidence in this experiment gives a clue as to which reactant is the limiting reagent?

3.   Why is the HCl added to the solution?

4.   Why are the exact amounts of water and HCl relatively unimportant?

5.   Give two sources of error as to why the percent yield in this experiment may be greater than 100%.
      1.

      2.

6.   Give two sources of error as to why the percent yield in this experiment may be less than 100%.
      1.

      2.

7.   Assuming the same amount of copper (II) chloride dihydrate as used in this experiment was used in another experiment, what is the maximum mass of aluminum that would make aluminum the limiting reagent?

8.   What physical evidence would show that aluminum is the limiting reagent?

9.   In industry, what would be a factor in making one reactant the limiting reagent and another reactant the excess reagent?

EXPERIMENT 4 LIMITING REAGENTS

Name _________________________

HOMEWORK EXERCISES
There are four pages of this homework.

1. 17.98 g of calcium chloride and 18.05 g of potassium hydroxide are added to 1.000 L of water and react to form calcium hydroxide and potassium chloride.

    a. Write the balanced molecular equation for this reaction. Include states.

    b. Write the ionic and net ionic equations for this reaction. Include states.

Ionic:

Net ionic:

    c. Determine the limiting reagent and how many grams of calcium hydroxide should be formed.

    d. How many grams of excess reagent are left over?

    e. If 10.56 g of calcium hydroxide are obtained in an experiment, what is the % yield of the experiment?

2. 25.64 g of copper (II) sulfate pentahydrate and 12.69 g of sodium phosphate are added to 1.000 L of water.

a. Predict the products and write the balanced molecular equation for this reaction. Include states.

b. Write the ionic and net ionic equations for this reaction. Include states.

Ionic:

Net ionic:

c. Determine the limiting reagent and how many grams of the precipitate are formed.

d. How many grams of excess reagent are left over?

e. If 9.95 g of the product were recovered, what is the % yield of the reaction?

3. 200.0 mL of 0.350 M lead (II) nitrate are added to 300.0 mL of 0.250 M sodium iodide.
    a. Predict the products and write the balanced molecular equation for this reaction. Include states.

    b. Write the ionic and net ionic equations for this reaction. Include states.

Ionic:

Net Ionic:

    c. Determine the limiting reagent and how many grams of the precipitate are formed.

    d. How many grams of the excess reagent are left over?

    e. What is the concentration (M) of the excess reagent in solution?

4. In the industrial synthesis of aspirin, 200 kg of salicylic acid, $C_7H_6O_3$, and 150 kg of acetic anhydride, $(CH_3CO)_2O$, form acetylsalicylic acid, $C_9H_8O_4$, (aspirin) and acetic acid, $C_2H_4O_2$.
    a. Write the balanced chemical reaction for this reaction.

    b. In a manufacturing process, 200 kg of aspirin are produced from 200 kg of impure salicylic acid and 150 kg of acetic anhydride. The salicylic acid is the limiting reagent. Calculate the purity of the salicylic acid.

    c. How many kg of excess reagent are left over, assuming there are no side reactions between the impurities in the limiting reagent and the excess reagent?

EXPERIMENT 5
POLYATOMIC IONS, AQUEOUS SOLUBILITY RULES,
NET IONIC EQUATIONS AND OXIDATION STATES

A variety of ionic compounds, including many containing polyatomic ions, will be mixed to demonstrate solubility rules. Molecular, ionic, and net ionic equations will be written for many of the reactions.

Key Chemical Reactions: See data sheet

Key Mathematical Equations: None

Foundational Chemistry Concepts: Polyatomic Ions
     Mastery of Table 5.1 is required
    Solubility Rules for Common Ionic Compounds in Water
     Mastery of Table 5.3 is required
    Writing molecular, ionic, and net ionic equations

Discussion

An ionic compound is composed of two different ions: a cation and an anion. A cation is an ion which has a positive charge because it has lost one or more electrons, such as Na^+, Ca^{2+}, Fe^{2+}, Al^{3+}, and ions from group 1A and 2A elements. An anion is an ion which has a negative charge because it has gained one or more electrons, such as Cl^-, O^{2-}, and most non-metals. Polyatomic ions are ion comprised of two or more elements, as shown in Table 5.1.

Table 5.1 Polyatomic Ions

+1	−1	−2	−3
NH_4^+ (ammonium)	CN^- (cyanide)	O_2^{-2} (peroxide)	
H_3O^+ (hydronium)	OCN^- (cyanate)	$C_2O_4^{-2}$ (oxalate)	
Hg_2^{2+} (mercury (I))	SCN^- (thiocyanate)		
	OH^- (hydroxide)	CrO_4^{-2} (chromate)	
	$C_2H_3O_2^-$ (acetate)	$Cr_2O_7^{-2}$ (dichromate)	
	MnO_4^- (permanganate)		
	NO_3^- (nitrate)		
	NO_2^- (nitrite)		
	$H_2PO_4^-$ (dihydrogen phosphate)	HPO_4^{-2} (hydrogen phosphate)	PO_4^{-3} (phosphate)
	HCO_3^- (bicarbonate)	CO_3^{-2} (carbonate)	AsO_4^{-3} (arsenate)
	HSO_4^- (hydrogen sulfate	SO_4^{-2} (sulfate)	
	HSO_3^- (hydrogen sulfite)	SO_3^{-2} (sulfite)	
	ClO_4^- (perchlorate) *	$S_2O_3^{-2}$ (thiosulfate)	
	ClO_3^- (chlorate) *		
	ClO_2^- (chlorite) *		
	ClO^- (hypochlorite) *		
	* bromo, iodo relatives		*Courtesy: Dr. Patty Wisian-Neilson*

61

In order for an ionic compound to be electrically neutral, the sum of the positive charges must equal the sum of the negative charges. This is done by multiplying one or both of the ions (or polyatomic ions) by small whole numbers. Most often, the subscript of the cation is equal to the charge of the anion and vice versa. Use parenthesis only on polyatomic ions if more than one polyatomic ion is required. Examples are shown in Table 5.2.

Table 5.2 Compounds Formed by Various Cations and Anions

Cation	Anion	Ionic Compound
Na^+	Cl^-	$NaCl$
Na^+	S^{2-}	Na_2S
Ca^{2+}	O^{2-}	CaO
Ca^{2+}	Cl^-	$CaCl_2$
Al^{3+}	Cl^-	$AlCl_3$
Al^{3+}	O^{2-}	Al_2O_3
Ag^+ **(Ag only carries a +1 charge)**	NO_3^-	$AgNO_3$
Na^+	$S_2O_3^{2-}$	$Na_2S_2O_3$
Al^{3+}	SO_4^{2-}	$Al_2(SO_4)_3$
Ca^{2+}	MnO_4^-	$Ca(MnO_4)_2$

When mixing ionic compounds in solution, a reaction takes place when a precipitate, gas, or liquid forms. Predicting a precipitate can be done using a solubility chart, shown in Table 5.3.

Table 5.3
Solubility Rules for Common Ionic Compounds in Water

Soluble	Insoluble

Group I and NH_4^+
 Nitrates (NO_3^-)
 Acetates (CH_3COO^-)
 Bicarbonates (HCO_3^-)
 Chlorates and Perchlorates
   (ClO_3^- and ClO_4^-)
 Halides (Cl^-, Br^-, I^-)
   Except Ag^+, Hg_2^{2+}, Pb^{2+}
 Sulfates (SO_4^{2-})
   Except Ag^+, Ca^{2+}, Sr^{2+}, Ba^{2+}, Hg_2^{2+}, Pb^{2+}
	Carbonates (CO_3^{2-})
	 Phosphates (PO_4^{3-})
	 Chromates (CrO_4^{2-})
	 Sulfides (S^{2-})
	
	 Except Group I and NH_4^+
	
	 Hydroxides (OH^-)
	 Except Group I, NH_4^+
	 and Ba^{2+}

For example, suppose an aqueous solution of $Ca(NO_3)_2$ is combined with an aqueous solution of Na_2SO_4. Since this is a double displacement reaction, switch the positive and negative ions or polyatomic ions from each compound and then balance the new molecule, making sure the charges are balanced. Finally, balance the whole equation. In this example,

$$Ca(NO_3)_2 + Na_2SO_4 \rightarrow CaSO_4 + 2\,NaNO_3$$

To see if a precipitate is formed, check the components with the solubility table. Since nitrates and compounds that contain elements from group 1A are all soluble, these species will remain in solution and will have the (aq) subscript for "aqueous". $CaSO_4$, however, is insoluble, and is therefore a precipitate (s) for "solid". The chemical reaction with states is as follows:

$$Ca(NO_3)_{2\,(aq)} + Na_2SO_{4\,(aq)} \rightarrow CaSO_{4\,(s)} + 2\,NaNO_{3\,(aq)}$$

Molecular, Ionic and Net Ionic Equations
A molecular equation shows all compounds in a balanced chemical reaction with complete, neutral molecular formulas. An ionic equation shows this reaction with all soluble ionic compounds dissociated into their ionic components. A net ionic equation shows the chemical equation with only the ions involved in the reaction. The ions not involved in the reaction, called spectator ions, are omitted. A simple procedure for this is as follows:

1. To write the balanced chemical equation, treat the reaction as a double displacement reaction: Switch the ions of the reactants to form molecular compounds in the products. Balance the compounds and the whole reaction.   ⎤ Molecular
2. Check the solubility of each reactant and product. Write the appropriate subscript for each compound. ⎦
3. If all compounds are soluble (aq subscript), no reaction occurs. Write no reaction or NR for the ionic and net ionic reactions.   ⎱ No Reaction
4. On the product side, break up all soluble (aq) components into their ionic forms.
    a. Subscripts that are not part of a polyatomic ion become stoichiometric multipliers to the ion.
        $$Ca(NO_3)_2 \text{ becomes } Ca^{2+} + 2\,NO_3^-$$ ⎱ Ionic
    b. Keep compounds that are liquids (H_2O), gases, and solids together. Do not dissociate them.
5. To obtain the net ionic equation, cancel identical species that appear on both sides of the equation. The cancelled species are called spectator ions.   ⎱ Net Ionic

Continuing the above example:

Molecular: $Ca(NO_3)_{2\,(aq)} + Na_2SO_{4\,(aq)} \rightarrow CaSO_{4\,(s)} + 2\,NaNO_{3\,(aq)}$

Ionic: $Ca^{2+}_{\,(aq)} + 2\,NO_3^-{}_{\,(aq)} + 2\,Na^+{}_{\,(aq)} + SO_4^{2-}{}_{\,(aq)} \rightarrow CaSO_{4\,(s)} + 2\,Na^+{}_{\,(aq)} + 2\,NO_3^-{}_{\,(aq)}$

Net ionic: $Ca^{2+}_{\,(aq)} + SO_4^{2-}{}_{\,(aq)} \rightarrow CaSO_{4\,(s)}$

If no precipitate, gas, or liquid forms, there is no reaction, and all components remain in their ionic form. For example, when $NaNO_3$ is mixed with KOH, no reaction occurs since all components remain in solution:

Molecular: $NaNO_{3\ (aq)} + KOH_{(aq)} \rightarrow NaOH_{(aq)} + KNO_{3\ (aq)}$

Ionic: $Na^+_{\ (aq)} + NO_3^-_{\ (aq)} + K^+_{\ (aq)} + OH^-_{\ (aq)} \rightarrow$ No reaction

Reactions with Liquid or Gas Formation

There are a number of different types of reactions (all involve either an acid or a base) that form a liquid or a gas. These reactions will not be performed in this experiment, as liquids and gases are not as readily seen as are precipitates. The molecular form of each type of these equations is shown.

- Acids react with metals (Group I, II, transition metals) to produce $H_{2\ (g)}$

$$H_2SO_{4\ (aq)} + Zn_{(s)} \rightarrow ZnSO_{4\ (aq)} + H_{2\ (g)}$$

- Acids react with carbonates and bicarbonates to produce $CO_{2\ (g)}$ and $H_2O_{(l)}$

$$H_2SO_{4\ (aq)} + MgCO_{3\ (s)} \rightarrow MgSO_{4\ (aq)} + H_2O_{(l)} + CO_{2\ (g)}$$

$$H_2SO_{4\ (aq)} + 2\ NaHCO_{3\ (s)} \rightarrow Na_2SO_{4\ (aq)} + 2\ H_2O_{(l)} + 2\ CO_{2\ (g)}$$

- Acids react with sulfites to produce $SO_{2\ (g)}$ and $H_2O_{(l)}$

$$2\ HNO_{3\ (aq)} + MgSO_{3\ (s)} \rightarrow Mg(NO_3)_{2\ (aq)} + H_2O_{(l)} + SO_{2\ (g)}$$

- Acids react with sulfides to produce $H_2S_{(g)}$

$$2\ HBr_{(aq)} + CaS_{(s)} \rightarrow CaBr_{2\ (aq)} + H_2S_{(g)}$$

- Acids react with bases to produce $H_2O_{(l)}$

$$H_2SO_{4\ (aq)} + 2\ NaOH_{(aq)} \rightarrow Na_2SO_{4\ (aq)} + 2\ H_2O_{(l)}$$

- Bases react with ammonium compounds to produce $NH_{3\ (g)}$ and $H_2O_{(l)}$

$$Ba(OH)_{2\ (aq)} + 2\ NH_4Cl_{(aq)} \rightarrow BaCl_{2\ (aq)} + 2\ NH_{3\ (g)} + 2\ H_2O_{(l)}$$

Oxidation States

The oxidation state of an atom within a molecule or ionic compound is the number of electrons gained or lost if the electrons were transferred to other atoms in the compound. Determining oxidation states of elements helps to construct the formula of a compound, as well as knowing if the element is being oxidized or reduced in an oxidation-reduction reaction. There are general guidelines in determining the oxidation of elements within a compound, shown in Table 5.4.

Table 5.4 General Guidelines for Assigning Oxidation Numbers

	Oxidation State	Exceptions
Diatomic Elements (ONLY Br_2, I_2, N_2, CL_2, H_2, O_2, F_2)	0	
Group 1A	+1	
Group 2A	+2	
Group 3A	+3	
Group 6A	Usually −2	
Group 7A	Usually −1	Cl and Br can vary
Group 8A	0	
Oxygen	−2	−1 in peroxides (H_2O_2) − ½ in superoxides (O_2^-)
Hydrogen	+1	−1 when bonded to metals (Li, Na, etc.)
Metals by themselves (e.g. Cu, S_8)	0	
Transition metals	All positive	

The sum of the oxidation states of the cations and the anions is zero in a neutral molecule and the charge of the polyatomic ion in a polyatomic ion.

For example, calculate the oxidation state of S in the polyatomic ion SO_4^{2-}. The sum of the oxidation state of S plus four times the oxidation state of O must add to −2, the charge of the polyatomic ion. It is known that oxygen is almost always −2.

$$(\_?\_ \times 1 \text{ atom of S}) + (-2 \times 4 \text{ atoms of O}) = -2$$

Using simple algebra, $(1 \times S) + -8 = -2$, so the oxidation state of S in SO_4^{2-} is +6.

Calculate the oxidation state of the following underlined elements:

1. $Na\underline{N}O_3$: $(+1 \times 1 \text{ atom of Na}) + (\_?\_ \times 1 \text{ atom of N}) + (-2 \times 3 \text{ atoms of O}) = 0$

   $N = +5$

2. $Ca(\underline{Mn}O_4)_2$: $(+2 \times 1 \text{ atom of Ca}) + (\_?\_ \times 2 \text{ atoms of Mn}) + (-2 \times 8 \text{ atoms of O}) = 0$

   $Mn = +7$ *Be sure to distribute the "2" subscript of the permanganate ion to all atoms in the polyatomic ion.*

3. $(NH_4)_2\underline{Cr}_2O_7$: $(+1 \times 2 \text{ ions of } NH_4^+) + (\_?\_ \times 2 \text{ atoms of Cr}) + (-2 \times 7 \text{ atoms of O}) = 0$

   $Cr = +6$

Prelab Note

Prelab should contain the following sections:

- Purpose
- Method
- Chemicals
- Waste
- Solubility Rules Chart, Table 5.3.

Chemicals

Ammonium carbonate, 0.2 M	Silver nitrate, 0.2 M
Barium chloride, 0.2 M	Sodium bicarbonate, 0.2 M
Copper (II) sulfate, 0.2 M	Sodium hydroxide, 0.2 M
Lead (II) acetate, 0.2 M	Sodium phosphate, 0.2 M
Potassium chromate, 0.2 M	Sodium sulfide, 0.2 M

Waste

Dispose of waste in the provided waste container. Rinse the well plates with DI water to the same container. Wipe each well thoroughly with a swab to remove all traces of color, solution, or precipitate (the swab can be thrown in the trash). Dry thoroughly.

Procedure

A. Initial Data Table

1. Add 2 drops of each solution (as directed in the data table) to a well in the well plate.

2. Record observations in each box as follows:
   - If a precipitate forms, write the chemical formula of the precipitate.
   - If no reaction takes place, write NR.
   - If a precipitate forms where no precipitate should form, write "Unknown".

B. Molecular, Ionic, and Net Ionic Reactions

For the marked boxes in the data table, write molecular, ionic, and net ionic reactions. Include states.

C. Unknowns

For each unknown provided, there is a possible ion in solution, written on the vial. Using solubility rules, suggest a test that could be performed to confirm that the ion is or is not present, and then carry out the test in the well plates.

> *Important Note*:
> For Ca^{2+}, do not use either $CuSO_4$ or K_2CrO_4 for testing. The compounds formed do not have significant enough K_{sp} values (a concept that will be discussed in depth later in this manual) to give a definitive test.

Disposal and Clean Up

Dispose of the contents of the well plates to the provided waste container. Rinse the well plates with DI water to the same container. Wipe each well thoroughly with a swab to remove all traces of color, solution, or precipitate (the swab can be thrown in the trash). Dry thoroughly.

Name ________________________

A. Initial Data Table

	A	B	C	D	E	F
1	Ammonium carbonate + Barium chloride 1	Barium chloride + Lead (II) acetate	Lead (II) acetate + Copper (II) sulfate 5	Copper (II) sulfate + Potassium chromate	Potassium chromate + Silver nitrate 8	Silver nitrate + Sodium bicarbonate
2	Ammonium carbonate + Lead (II) acetate	Barium chloride + Copper (II) sulfate 2	Lead (II) acetate + Potassium chromate	Copper (II) sulfate + Sodium bicarbonate	Potassium chromate + Sodium bicarbonate	Silver nitrate + Sodium hydroxide
3	Ammonium carbonate + Copper (II) sulfate	Barium chloride + Potassium chromate	Lead (II) acetate + Silver nitrate	Copper (II) sulfate + Sodium hydroxide 7	Potassium chromate + Sodium hydroxide	Silver nitrate + Sodium phosphate 9
4	Ammonium carbonate + Potassium chromate	Barium chloride + Silver nitrate	Lead (II) acetate + Sodium bicarbonate	Copper (II) sulfate + Sodium phosphate	Potassium chromate + Sodium phosphate	Silver nitrate + Sodium sulfide
5	Ammonium carbonate + Silver nitrate	Barium chloride + Sodium bicarbonate	Lead (II) acetate + Sodium hydroxide	Copper (II) sulfate + Sodium sulfide	Potassium chromate + Sodium sulfide	Unknown #1
6	Ammonium carbonate + Sodium bicarbonate	Barium chloride + Sodium hydroxide 3	Lead (II) acetate + Sodium phosphate	Sodium phosphate + Sodium sulfide	Sodium bicarbonate + Sodium hydroxide	Unknown #2
7	Ammonium carbonate + Sodium hydroxide	Barium chloride + Sodium phosphate 4	Lead (II) acetate + Sodium sulfide 6	Sodium hydroxide + Sodium phosphate	Sodium bicarbonate + Sodium phosphate	Unknown #3
8	Ammonium carbonate + Sodium phosphate	Barium chloride + Sodium sulfide	Ammonium carbonate + Sodium sulfide	Sodium hydroxide + Sodium sulfide	Sodium bicarbonate + Sodium sulfide	Unknown #4

B. Molecular, Ionic, and Net Ionic Equations (Include states in all reactions)

1	Molecular:
	Ionic:
	Net Ionic:
2	Molecular:
	Ionic:
	Net Ionic:
3	Molecular:
	Ionic:
	Net Ionic:
4	Molecular:
	Ionic:
	Net Ionic:
5	Molecular:
	Ionic:
	Net Ionic:
6	Molecular:
	Ionic:
	Net Ionic:
7	Molecular:
	Ionic:
	Net Ionic:
8	Molecular:
	Ionic:
	Net Ionic:
9	Molecular:
	Ionic:
	Net Ionic:

Write expected equations, regardless of what was observed

C. Unknowns

Unknown number	Possible ion present	Is this ion present?

What reagent could be added to confirm or deny the presence of this ion? _______________

What is the chemical formula of the precipitate that would occur? _______________________

Write the net ionic equation of this reaction, <u>assuming a precipitate occurred</u> (include states):

Unknown number	Possible ion present	Is this ion present?

What reagent could be added to confirm or deny the presence of this ion? _______________

What is the chemical formula of the precipitate that would occur? _______________________

Write the net ionic equation of this reaction, <u>assuming a precipitate occurred</u> (include states):

Unknown number	Possible ion present	Is this ion present?

What reagent could be added to confirm or deny the presence of this ion? _______________

What is the chemical formula of the precipitate that would occur? _______________________

Write the net ionic equation of this reaction, <u>assuming a precipitate occurred</u> (include states):

Unknown number	Possible ion present	Is this ion present?

What reagent could be added to confirm or deny the presence of this ion? _______________

What is the chemical formula of the precipitate that would occur? _______________________

Write the net ionic equation of this reaction, <u>assuming a precipitate occurred</u> (include states):

<u>HOMEWORK EXERCISES</u>   Name_____________________

This homework is 4 pages: correct and balanced chemical formulas must be written.

1. Give an example (chemical formula) of the following in an aqueous solution. If no compound exists that meets the criteria, write "none". All chemical formulas must be balanced.

Insoluble halide _______________   Soluble halide _______________

Insoluble sulfate _______________   Soluble sulfate _______________

Insoluble nitrate _______________   Soluble sulfide _______________

Insoluble hydroxide _____________   Soluble hydroxide ____________

Insoluble phosphate ____________   Soluble acetate _______________

2. Write formulas of compounds that meet the following:
    Note: Both answers in each part should contain the same cation.

   a. Contains a Group 2A cation that is

insoluble as a carbonate ___$MgCO_3$_____ and soluble as a sulfate ___$MgSO_4$_____

   b. Contains a cation that is

insoluble as a hydroxide _____________ and soluble as a sulfate _____________

   c. Contains a cation that is

soluble as a hydroxide _______________and insoluble as a sulfate ____________

   d. Contains a Group 2A cation that is

insoluble as a phosphate _____________ and soluble as a halide ____________

   e. Contains a +2 cation that is

insoluble as a chromate _____________ and insoluble as a halide ___________

   f. Contains a cation that is

soluble as a nitrate _______________ and soluble as a sulfide _____________

   g. Contains a transition metal cation that is

insoluble as a carbonate ____________ and insoluble as a phosphate __________

3. Write the molecular (M), ionic (I), and net ionic (NI) chemical equations for the following reactions in aqueous solution. <u>Use chemical formulas and state subscripts for all species</u>. If a reaction does not occur, complete the M and I and write "NR" in the NI.

1.	sodium chloride + silver nitrate
M	
I	
NI	

2.	lead (II) acetate + potassium phosphate
M	
I	
NI	

3.	aluminum iodide + barium hydroxide
M	
I	
NI	

4.	magnesium sulfate + calcium chloride
M	
I	
NI	

5.	sodium hydroxide + barium nitrate
M	
I	
NI	

6.	ammonium sulfide + mercury (I) acetate
M	
I	
NI	

7.	iron (III) sulfate + strontium bromide
M	
I	
NI	

8.	lead (II) bicarbonate + potassium chromate
M	
I	
NI	

9.	manganese (II) perchlorate + rubidium carbonate
M	
I	
NI	

10.	lithium dichromate + ammonium thiosulfate
M	
I	
NI	

4. Write the molecular (M) and net ionic (NI) chemical equations for the following liquid/gas formation reactions in aqueous solution. <u>Use chemical formulas and state subscripts for all species.</u> Refer to the section in this lab of "Reactions with Liquid or Gas Formation" to complete this section.

1.	nitric acid + sodium sulfide
M	
I	
NI	

2.	hydrochloric acid + calcium carbonate
M	
I	
NI	

3.	ammonium phosphate + lithium hydroxide
M	
I	
NI	

4.	acetic acid + potassium sulfite
M	
I	
NI	

5. Calculate the oxidation state of each underlined atom:

$K\underline{Cl}O_4$		$NaH\underline{S}O_3$	
$\underline{S}_2O_3^{2-}$		$\underline{Cl}O_2^{-}$	
$Ba_3(\underline{P}O_4)_2$		$\underline{Cr}_2O_7^{2-}$	
$\underline{N}O_3^{-}$		$\underline{Bi}_2(CrO_4)_3$	
$\underline{N}O_2^{-}$		$Ca\underline{C}_2O_4$	

EXPERIMENT 6
ACID-BASE TITRATIONS

The concept of titration is introduced by standardizing a solution of sodium hydroxide and then using this solution to determine the percent of acetic acid in a sample and the molar mass of an unknown solid acid.

Key Chemical Reactions:

$$HA_{(aq)} \; + \; NaOH_{(aq)} \; \rightarrow \; NaA_{(aq)} \; + \; H_2O_{(l)}$$

$$H_2A_{(aq)} \; + \; 2\,NaOH_{(aq)} \; \rightarrow \; Na_2A_{(aq)} \; + \; 2\,H_2O_{(l)}$$

Key Mathematical Equations:

$$1.0000 \text{ g KHP} \times \frac{1 \text{ mole KHP}}{204.23 \text{ g KHP}} \times \frac{1 \text{ mole NaOH}}{1 \text{ mole KHP}} \times \frac{1}{0.02000 \text{ L NaOH}} = 0.2448 \text{ M NaOH}$$

$$\frac{0.500 \text{ g HA}}{10.0 \text{ g solution}} \times 100 = 5.00 \text{ \% HA in solution}$$

$$0.01500 \text{ L NaOH} \times \frac{0.2448 \text{ moles NaOH}}{L} \times \frac{1 \text{ mole HA}}{1 \text{ mole NaOH}} = 3.672 \times 10^{-3} \text{ moles HA}$$

$$\text{molar mass of HA} = \frac{0.500 \text{ g HA}}{3.672 \times 10^{-3} \text{ moles HA}} = 136.2 \text{ g/mole HA}$$

<u>Discussion</u>

A solution of accurately known molarity is called a standard solution. A standard solution is used to "standardize" other solutions (determine their molarities), and can also be used to determine the number of moles of a reactant present in a solution of unknown concentration. A standard solution can be prepared from the pure form of a solid substance, carefully weighed, and then dissolved in a measured total volume of solution.

Some acid and base solutions, such as hydrochloric acid and sodium hydroxide, cannot be prepared by this method. HCl is a gas at room temperature and is difficult to weigh accurately, and solid NaOH absorbs both water and carbon dioxide on exposure to air. Their molarities must be determined by standardizing them against some other solid acid or base just prior to using them. Such a solid acid or base that is obtained in a high state of purity and which can be weighed accurately is called a primary standard. One of the most commonly used primary standard acids is potassium hydrogen phthalate, $KHC_8H_4O_4$ (abbreviated KHP). It is a monoprotic acid, obtained in a very pure and easily weighable form.

The process of a controlled addition of a solution of one reactant to a fixed quantity of another reactant to reach a desired endpoint is called titration. In neutralization reactions of acids and bases in aqueous solution, the reactions usually involve hydrogen ions provided by the acids and hydroxide ions provided by the bases.

If HA is used as the general formula for a monoprotic acid (one hydrogen is available to react), the neutralization reaction is as follows:

$$HA_{(aq)} + NaOH_{(aq)} \rightarrow NaA_{(aq)} + H_2O_{(l)}$$
$$\text{acid} \qquad \text{base} \qquad \text{salt} \qquad \text{water}$$

Therefore, if a monoprotic acid is titrated by the addition of a standardized solution of NaOH, the reaction is complete and the equivalence point is reached when the number of moles of NaOH added is equal to the number of moles of acid originally present. At this point, the only solute present in the solution is the salt formed in the reaction. The pH at the equivalence point depends on the nature of the salt formed.

Similarly, when a diprotic acid is titrated, two hydrogens are available to react and two moles of NaOH must be added for every one mole of the acid (H_2A) present to reach the equivalence point.

$$H_2A_{(aq)} + 2\,NaOH_{(aq)} \rightarrow Na_2A_{(aq)} + 2\,H_2O_{(l)}$$

The endpoint is usually signaled by the change in color of a dye, called an indicator. The indicator can exist in two forms, an acidic form and a basic form, which differ in color. Phenolphthalein, for example, is colorless in its acidic form and pink in its basic form. Since the equivalence point for different acid–base pairs can occur at different pH values, it is important in a titration to select an indicator which changes in color at a pH which matches the pH at the equivalence point of the acid-base pair involved in the titration. Phenolphthalein, for example, changes color at a pH of ~9. It would be a suitable indicator for an acetic acid–sodium hydroxide titration whose equivalence point is close to pH 9, but would not be satisfactory for a hydrochloric acid–ammonia titration, which has an equivalence point at pH ~5. All chemistry texts have sections on how to choose the proper indicator for a specific titration.

After standardizing the NaOH solution in this experiment, a measured volume of an acid solution will be used to determine the concentration of the acid in solution. Next, a sample of an unknown solid acid will be dissolved and titrated, and the molar mass of the acid will be determined.

Example
1. <u>Standardization of NaOH solution using KHC$_8$H$_4$O$_4$ (KHP)</u>:
A 0.9566 g sample of KHC$_8$H$_4$O$_4$ required 23.42 mL of NaOH solution to reach an equivalence point with phenolphthalein as an indicator. What is the molarity of the NaOH solution?

$$KHC_8H_4O_4 + NaOH \rightarrow NaKC_8H_4O_4 + H_2O$$

Since $KHC_8H_4O_4$ is monoprotic, one mole of $KHC_8H_4O_4$ will react with one mole of NaOH, and the equivalence point is reached when the number of moles of NaOH added are equal to the number of moles of KHP present in the solution.

$$0.9566 \text{ g KHP} \times \frac{1 \text{ mole KHP}}{204.23 \text{ g KHP}} \times \frac{1 \text{ mole NaOH}}{1 \text{ mole KHP}} \times \frac{1}{0.02342 \text{ L NaOH}} = 0.2000 \text{ M NaOH}$$

2. Calculation of the percent acid in a solution:
A 10.0 mL sample of an acetic acid solution is diluted with water. Two drops of phenolphthalein are added, and the solution is titrated by the addition of a 0.2000 M solution of NaOH. 53.60 mL of the NaOH solution were required to reach the equivalence point. Calculate the percent of acetic acid (HAc) in the sample. Assume the density of the solution is 1.00 g/mL.

$$HC_2H_3O_2 + NaOH \rightarrow NaC_2H_3O_2 + H_2O$$

Since acetic acid is a monoprotic acid, moles HAc = moles NaOH at the equivalence point. The molarity of the HAc is calculated as follows:

$$0.05360 \text{ L NaOH} \times \frac{0.2000 \text{ moles NaOH}}{L} \times \frac{1 \text{ mole HAc}}{1 \text{ mole NaOH}} \times \frac{1}{0.0100 \text{ L soln}} = 1.07 \text{ M HAc}$$

Using this molarity, the molar mass of acetic acid, and the density of the solution, the mass of the acetic acid per liter of solution and the percent acetic acid can be calculated.

$$1.07 \text{ M HAc} \times \frac{60.05 \text{ g HAc}}{1 \text{ mole HAc}} = 64.4 \text{ g HAc/L}$$

$$64.4 \text{ g HAc/L} \times 0.01000 \text{ L solution} = 0.644 \text{ g HAc in the sample}$$

$$10.0 \text{ mL solution} \times \frac{1.00 \text{ g solution}}{1 \text{ mL solution}} = 10.0 \text{ g solution}$$

$$\frac{0.644 \text{ g HAc}}{10.0 \text{ g solution}} \times 100 = 6.44 \text{ \% HAc in solution}$$

3. Determination of the molar mass of an unknown solid acid:
A 0.5200 g sample of a solid monoprotic acid (HA) is dissolved in deionized water. Two drops of phenolphthalein are added, and the solution is titrated by the addition of a 0.2000 M solution of NaOH. 17.33 mL of the NaOH solution were required to reach the equivalence point. Calculate the molar mass of the acid.

$$HA + NaOH \rightarrow NaA + H_2O$$

Since this is a monoprotic acid, moles HA = moles NaOH at the equivalence point, and the molar mass of the acid is calculated as follows:

$$0.01733 \text{ L NaOH} \times \frac{0.2000 \text{ moles NaOH}}{\text{L}} \times \frac{1 \text{ mole HA}}{1 \text{ mole NaOH}} = 3.466 \times 10^{-3} \text{ moles HA}$$

$$\text{molar mass of HA} = \frac{0.5200 \text{ g HA}}{3.466 \times 10^{-3} \text{ moles HA}} = 150.0 \text{ g/mole HA}$$

If the solid acid was a diprotic acid, the calculations would be as follows:

$$H_2A_{(aq)} \;+\; 2\,NaOH_{(aq)} \;\rightarrow\; Na_2A_{(aq)} \;+\; 2\,H_2O_{(l)}$$

$$0.01733 \text{ L NaOH} \times \frac{0.2000 \text{ moles NaOH}}{\text{L}} \times \frac{1 \text{ mole } H_2A}{2 \text{ moles NaOH}} = 1.733 \times 10^{-3} \text{ moles } H_2A$$

$$\text{molar mass of HA} = \frac{0.5200 \text{ g } H_2A}{1.733 \times 10^{-3} \text{ moles } H_2A} = 300.1 \text{ g/mole } H_2A$$

Chemicals
  Potassium hydrogen phthalate (KHP), $KHC_8H_4O_{4(s)}$
  Sodium hydroxide, $NaOH_{(aq)}$
  Acetic acid, $HC_2H_3O_{2(aq)}$
  Phenolphthalein indicator
  Various solid organic acids

Waste
  All waste can be disposed down the drain.

Procedure
A. Standardization of the NaOH solution
1. Obtain NaOH in a beaker and normalize a buret using 5–10 mL of the NaOH solution that is to be standardized. Fill the buret and drain enough of the solution to bring the meniscus down to the graduated scale.

2. Weigh out approximately 1.2 g (read to the nearest 0.0001 g) of the primary standard acid, $KHC_8H_4O_4$, KHP. KHP is a monoprotic acid with a molar mass of 204.23 g/mole.

3. Place the KHP in a clean 125 mL Erlenmeyer flask. Add about 50 mL of deionized (DI) water and 3 to 4 drops of phenolphthalein.

4. Record the initial volume of the NaOH solution in the buret, read to the nearest 0.01 mL, and then add the NaOH to the $KHC_8H_4O_4$ solution, at first in 1 mL increments, gradually slowing down to a few drops at a time as the equivalence point is approached. During the addition, swirl the acid solution but be careful to ensure that all the NaOH solution from the buret goes directly into the flask. The solution will

develop a pink color in the region where drops of the base enter the solution. As the titration nears the equivalence point, a longer period of mixing will be necessary for the color to disappear. When an extremely faint but definite pink color develops and is stable for at least thirty seconds, the equivalence point has been reached. During the titration, also wash down any droplets of liquid that may have splashed up onto the sides of the flask or were left on the outside tip of the buret with DI water.

5. Carefully touch the tip of the buret to the inside of the titration flask and make sure that this small amount of NaOH reaches the liquid in the flask. Rinse with DI water.

6. Allow 30 seconds for drainage along the inner walls of the buret, then read and record the final NaOH volume.

7. Make a second and third determination (if the first two molarities do not agree within 5%) in the same manner. Between titrations rinse the Erlenmeyer flask once with tap water and then once with deionized water. It is not necessary to dry the flask.

$$\frac{M_1 - M_2}{\left(\frac{M_1 + M_2}{2}\right)} \times 100 < \pm 5\%$$

B. Determination of percent acetic acid in a solution
1. Using an autodispenser, dispense exactly 10.0 mL of an acetic acid solution to a 125 mL Erlenmeyer flask using the following method: First, place a waste beaker under the delivery spout and pull the autodispenser barrel upwards. Touch any drops at the end of the delivery spout to the side of the waste beaker. Next, place the 125 mL Erlenmeyer flask under the spout and gently push the autodispenser barrel down. Continue with step 3. (If an autodispenser is not available, start with step 2.)

2. *Complete this step only if the acetic acid solution was not dispensed with an autodispenser.* Transfer about 50 mL of the acetic acid solution to a clean, dry, 100 mL beaker. Normalize a 10.00 mL pipet with the solution and then pipet 10.00 mL of the solution into the clean 125 mL Erlenmeyer flask. Let the pipet completely drain and then touch the tip of the pipet to the inside wall of the Erlenmeyer flask to complete the delivery of the 10.00 mL aliquot. Do not force out the remaining liquid.

3. Add approximately 25 mL of deionized water to the flask, followed by 3 or 4 drops of phenolphthalein.

4. Titrate the acetic acid solution with the standardized NaOH solution as outlined in Part A. Make a second and third determination (if the first two percents do not agree within 5%) in the same manner. Between titrations rinse the Erlenmeyer flask once with tap water and then once with deionized water. It is not necessary to dry the flask.

C. Determination of the molar mass of an acid

1. Weigh out between 0.4000 and 0.5000 g of the unknown acid and transfer it to the clean 125 mL Erlenmeyer flask.

2. Add 50 mL of deionized water to the flask followed by 3 or 4 drops of phenolphthalein. Use DI water to rinse down any solid particles of the unknown that have adhered to the neck or upper walls of the flask. If the solid is not completely in solution, the titration can still be started. The remaining particles will dissolve during the titration as the acid is converted to its more soluble sodium salt.

    Some of the unknown acids dissolve very slowly. These solutions may be warmed gently for two or three minutes on a hot plate and practically all of the solid should dissolve. Do not overheat. Contact your instructor before doing this.

3. Swirl the mixture gently for a minute or so and then titrate. If directed by the instructor, make a second determination of the molar mass.

4. After the titrations have been completed, carefully rinse the buret several times with deionized water until a test with pH paper shows a pH 7 or less. Strong alkalis such as NaOH etch glass slowly and in a short time will make an expensive buret worthless.

Important Points:
1. The amount of deionized water used for dilution or to dissolve a solid acid is (within reasonable limits) of no importance in these titration experiments. The number of moles of acid present in the container is not altered by the addition of pure water. It is determined solely by the grams of acid present. For example, once a specific volume of an acetic acid solution is delivered into a flask, deionized water may be added without altering the moles of acid in the flask.

2. The indicator should be added sparingly to the solution to be titrated. Since phenolphthalein is an acid, it will require some base to convert it to its basic form, the use of excessive amounts of indicator can lead to error.

3. It is always important to write the equation of the reaction to determine the stoichiometric ratio of the moles of acid to the moles of base.

EXPERIMENT 6   ACID-BASE TITRATIONS

LABORATORY REPORT SHEET

Name_______________________________________ Date________________

Partner_______________________________

A. Standardization of the NaOH Solution

**READ ALL EQUIPMENT TO THE
CORRECT PLACE VALUE AND
RECORD ALL VALUES WITH
PROPER UNITS!**

	Trial 1	Trial 2	Trial 3
Mass of $KHC_8H_4O_4$			
Initial buret reading			
Final buret reading			
Volume of NaOH added			

Calculation of NaOH molarity from Trial 1:

Calculation of NaOH molarity from Trial 2:

Calculation of NaOH molarity from Trial 3 (if necessary):

Average NaOH molarity ________________________________

Only include trials within ± 5% of each other

LABORATORY REPORT SHEET

B. Determination of the Percent Acetic Acid in a Solution

	Trial 1	Trial 2	Trial 3
Unknown #			
Volume of acetic acid solution			
Mass of acetic acid solution (Assume the density is that of water)			
Initial buret reading			
Final buret reading			
Volume of NaOH added			

1. Molarity of acetic acid solution:
   Trial 1

   Trial 2

   Trial 3 (if necessary)

        Average acetic acid solution molarity _______________________________

2. Mass of acetic acid per liter of solution:

3. Percent of acetic acid in solution:

LABORATORY REPORT SHEET

Name____ Date________________

C. Determination of the Molar Mass of a Solid Acid

Write the generic reaction for a titration of a monoprotic acid with NaOH:

▶ ____

Write the generic reaction for a titration of a diprotic acid with NaOH:

▶ ____

Unknown # _______________________

	Trial 1	Trial 2 (only if needed)
Mass of solid acid		
Initial buret reading		
Final buret reading		
Volume of NaOH		

Moles of unknown acid:

Molar mass of unknown acid:

Error Analysis

1. If the KHP was overtitrated with NaOH, the molarity of the NaOH would appear

   _________________.

2. If not all of the KHP was dissolved and the titration was stopped, the molarity of the NaOH would appear _________________.

3. If the weighing paper was not tared in the NaOH determination, the molarity of the NaOH would appear _________________.

4. If the acetic acid solution was overtitrated with NaOH, the acetic acid molarity and % would appear _________________.

5. If the unknown solid acid was overtitrated with NaOH, the molar mass of the acid would appear _________________.

6. If the unknown solid acid was not dissolved and the titration was stopped, the molar mass of the acid would appear _________________.

Name _______________________________

HOMEWORK EXERCISES

1. Determine the molarity of a NaOH solution if 0.8427 g of KHP are required to standardize 15.57 mL of NaOH.

2. A 0.5192 g sample of a solid monoprotic acid (molar mass of 180.0 g/mole) was dissolved in water and titrated by the addition of an aqueous solution of NaOH. To reach the equivalence point, 15.96 mL of NaOH solution were required.
   a. Calculate the molarity of the NaOH solution.

   b. Write the generic representation of the reaction of a monoprotic acid with NaOH. Then, calculate the molar mass of an unknown monoprotic acid if 15.64 mL of the NaOH solution from (a) was required to titrate 0.8256 g of the acid.

   Reaction: ____

   c. Write the generic representation of the reaction of a diprotic acid with NaOH. Then, calculate the molar mass of an unknown diprotic acid if 15.64 mL of the NaOH solution from (a) was required to titrate 0.8256 g of the acid.

   Reaction: ____

Continued

3. Barium hydroxide reacts with phosphoric acid to form barium phosphate and water.
   a. Write the balanced equation for this reaction. Include states.

   b. How many mL of 0.2195 M barium hydroxide would be required to titrate 12.57 mL of phosphoric acid. The density of phosphoric acid is 1.834 g/mL.

4. A 10.00 mL sample of a solution of acetic acid in water was titrated to the endpoint by the addition of 15.89 mL of 0.2432 M NaOH.
   a. Calculate the molarity of the acetic acid solution.

   b. Calculate the grams of acetic acid present per liter of the unknown solution.

   c. Calculate the percent acetic acid in the solution. Assume the density of the solution is 1.00 g/mL.

   d. Calculate the volume of this NaOH solution that would be required to titrate 50.00 mL of vinegar. Vinegar is 5.00 percent (by mass) acetic acid. Assume that the density of the vinegar solution is the same as that of water.

EXPERIMENT 7
OXIDATION-REDUCTION TITRATIONS

Using titration, two electron transfer (redox) reactions in aqueous solution will be studied. In the first titration, a solution containing the oxidizing agent potassium permanganate, $KMnO_4$, will be standardized by reacting it with a known mass of a primary standard reducing agent, $Fe(NH_4)_2(SO_4)_2 \cdot 6\ H_2O$, also known as Mohr's salt. In the second titration, the standardized $KMnO_4$ solution will be used to determine the percent of hydrogen peroxide in an unknown aqueous solution.

Key Chemical Reactions:

$$5\ Fe^{2+}_{(aq)} + MnO_4^{-}{}_{(aq)} + 8\ H^{+}_{(aq)} \rightarrow 5\ Fe^{3+}_{(aq)} + Mn^{2+}_{(aq)} + 4\ H_2O_{(l)}$$

$$2\ MnO_4^{-}{}_{(aq)} + 5\ H_2O_{2(aq)} + 6\ H^{+}_{(aq)} \rightarrow 2\ Mn^{2+}_{(aq)} + 5\ O_{2(g)} + 8\ H_2O_{(l)}$$

Key Mathematical Equations:

$$g\ Mohr's\ Salt \times \frac{1\ mole\ Mohr's\ Salt}{392.1\ g} \times \frac{1\ mole\ KMnO_4}{5\ moles\ Mohr's\ Salt} \times \frac{1}{L\ KMnO_4} = M_{KMnO_4}$$

$$M_{KMnO_4} \times L\ of\ KMnO_4\ added \times \frac{5\ moles\ H_2O_2}{2\ moles\ KMnO_4} \times \frac{34.02\ g\ H_2O_2}{1\ mole\ H_2O_2} = g\ H_2O_2$$

$$\%\ H_2O_2 = \frac{grams\ of\ H_2O_2}{grams\ of\ sample} \times 100$$

<u>Discussion</u>

There are four important points of oxidation-reduction reactions.

1. Reactions in which electrons are transferred are known as <u>oxidation–reduction</u> or <u>redox reactions</u>.
2. <u>Oxidation</u> involves the loss of electrons; <u>reduction</u> involves the gain of electrons.
3. The reactant containing the atom that gains electrons (and is reduced) is an <u>oxidizing agent</u>; the reactant containing the atom which loses electrons (and is oxidized) is a <u>reducing agent</u>.
4. In any redox reaction, for complete reaction, the number of electrons gained by the total moles of oxidizing agent must be stoichiometrically equal to the number lost by the total moles of the reducing agent. The point at which this occurs in a titration is called the <u>equivalence point</u>.

In the experiment "Acid-Base Titrations", <u>neutralization</u> titrations were carried out by the addition of solutions of bases to acids. Since many redox reactions take place in aqueous solution, it is also possible to carry out a <u>redox</u> titration by the addition of a solution of an oxidizing agent to a reducing agent – or vice versa. The equivalence point will be reached when the number of electrons gained by the moles of the oxidizing agent is equal the number of electrons given up by the moles of the reducing agent.

Standardization of KMnO₄

Potassium permanganate, $KMnO_4$, is one of the most commonly used laboratory oxidizing agents (Mn^{7+} is reduced, by gaining electrons to become Mn^{2+}). The permanganate ion, MnO_4^-, reacts as follows in an acidic solution (ionic form):

$$MnO_4^-{}_{(aq)} \; + \; 8\,H^+{}_{(aq)} \; + \; 5\,e^- \; \rightarrow \; Mn^{2+}{}_{(aq)} \; + \; 4\,H_2O_{(l)}$$

In this titration, as the purple MnO_4^- solution is added from a buret to the solution of a reducing agent in a flask, the colorless Mn^{2+} is formed and the solution in the flask remains virtually colorless. However, once the equivalence point is reached, any additional MnO_4^- added will not be reduced and a permanent faint pink color will appear in the flask, signaling the attainment of the endpoint. The MnO_4^- acts as its own indicator in the redox titration.

$KMnO_4$ solutions are somewhat unstable and will change concentration with time and exposure to light or air. Therefore the concentration of the $KMnO_4$ solutions must be determined just before use by standardizing it using the primary standard reducing agent $Fe(NH_4)_2(SO_4)_2 \cdot 6\,H_2O$ (Mohr's Salt, molar mass = 392.1 g/mole). The $KMnO_4$ standardization is carried out by carefully weighing out a sample of the Mohr's salt, dissolving it in an acid solution, and then titrating it with the $KMnO_4$ solution. The following equation gives the ionic half reaction for the Fe^{2+} ion in Mohr's salt.

$$Fe^{2+}{}_{(aq)} \; \rightarrow \; Fe^{3+}{}_{(aq)} \; + \; e^-$$

The above Fe^{2+} ionic half reaction is multiplied through by 5 in order to cancel out electrons on both sides of the net ionic reaction when the two half reactions are combined. Electrons do not appear in the balanced net ionic reaction. Combining these two ionic half reaction equations, the balanced net ionic reaction for the standardization of $KMnO_4$ with Mohr's salt is

$$5\,Fe^{2+}{}_{(aq)} \; + MnO_4^-{}_{(aq)} \; + \; 8\,H^+{}_{(aq)} \; \rightarrow \; 5\,Fe^{3+}{}_{(aq)} \; + \; Mn^{2+}{}_{(aq)} \; + \; 4\,H_2O_{(l)}$$

Using this balanced chemical equation, coupled with the mass of the Mohr's salt and the volume of $KMnO_4$ added, the concentration of the $KMnO_4$ can be calculated.

$$\text{g Mohr's Salt} \times \frac{1\ \text{mole Mohr's Salt}}{392.1\ \text{g}} \times \frac{1\ \text{mole KMnO}_4}{5\ \text{moles Mohr's Salt}} \times \frac{1}{\text{L KMnO}_4} = M_{KMnO_4}$$

Percent H₂O₂

The standardized $KMnO_4$ will be used to find the percent hydrogen peroxide, H_2O_2, in an unknown solution. Hydrogen peroxide is oxidized by the permanganate ion in acid solution according to the following half reactions:

$$MnO_4^-{}_{(aq)} \; + \; 8\,H^+{}_{(aq)} \; + \; 5\,e^- \; \rightarrow \; Mn^{2+}{}_{(aq)} \; + \; 4\,H_2O_{(l)} \qquad \text{(reduction)}$$

$$H_2O_2{}_{(aq)} \; \rightarrow \; O_2{}_{(g)} + \; 2\,H^+{}_{(aq)} + \; 2\,e^- \qquad \text{(oxidation)}$$

Multiplying the top equation by 2 and the bottom equation by 5 (so the electrons cancel each other out), the balanced net ionic reaction becomes

$$2\ MnO_4^{-}{}_{(aq)} + 5\ H_2O_{2(aq)} + 6\ H^{+}{}_{(aq)} \rightarrow 2\ Mn^{2+}{}_{(aq)} + 5\ O_{2(g)} + 8\ H_2O_{(l)}$$

The calculation of the percent H_2O_2 is a two step process as follows:

1. Calculate the grams of H_2O_2:

$$M_{KMnO_4} \times L\ of\ KMnO_4\ added \times \frac{5\ moles\ H_2O_2}{2\ moles\ KMnO_4} \times \frac{34.02\ g\ H_2O_2}{1\ mole\ H_2O_2} = g\ H_2O_2$$

2. Calculate the percent H_2O_2:

$$\%\ H_2O_2 = \frac{grams\ of\ H_2O_2}{grams\ of\ sample} \times 100$$

Chemicals
   Aqueous potassium permanganate, $KMnO_{4(aq)}$
   Sulfuric acid, 6 M, $H_2SO_{4(aq)}$
   Mohr's salt, $Fe(NH_4)_2(SO_4)_2 \cdot 6\ H_2O_{(s)}$
   Hydrogen peroxide, $H_2O_{2(aq)}$

Waste
   All waste can be disposed down the drain.

Procedure

Standardization of the KMnO₄ solution
1. Obtain 80 mL of the $KMnO_4$ solution in a clean and dry 100 mL beaker, and normalize the buret with the $KMnO_4$ solution.

2. Tare a weighing paper and weigh between 0.6 and 0.8 g of Mohr's Salt.

3. Transfer the salt to a 250 mL Erlenmeyer flask. Add about 40 mL of DI water and approximately 10 mL of 6 M H_2SO_4. Wash down any particles of salt on the sides of the flask with DI water. Swirl to dissolve the salt, making sure all the salt dissolves.

4. Titrate the salt with the $KMnO_4$ solution to the faintest pink color detectable, recording the initial and final buret readings. Repeat this titration until two $KMnO_4$ molarities agree within 5% of each other.

$$\frac{M_1 - M_2}{\left(\frac{M_1 + M_2}{2}\right)} \times 100 < \pm 5\%$$

Determination of peroxide by permanganate
NOTE: Work accurately but swiftly during this titration. The peroxide solution tends to decompose upon exposure to air and light.

1. Weigh a clean and dry 125 mL Erlenmeyer flask.

2. Add approximately 10 g (about 10 mL) of the unknown peroxide solution to the Erlenmeyer flask using a clean disposable pipet and reweigh.

3. Add about 30 mL of DI water and 10 mL of 6 M sulfuric acid.

4. Begin the titration by adding about one mL of permanganate from the buret while constantly swirling the flask. The initial reaction may be slow, and several minutes may be required for this first portion of permanganate to react completely (this is infrequent, however). The reaction is catalyzed by Mn^{2+}, one of the products formed, and from this point the reaction proceeds more rapidly.
5. Continue adding permanganate until the color begins to persist. Finally, add the permanganate dropwise with continued mixing until a pink color persists for at least one minute. Read and record the final volume.

6. Repeat the titration. Discard the remainder of the permanganate solution immediately after the second titration, and rinse the buret and the tip at least twice with DI water.

7. Clean all glassware, including the buret, thoroughly so that no traces of $KMnO_4$ are present.

LABORATORY REPORT SHEET

Name________________________________   Date__________________

Partner________________________________

STANDARDIZATION OF $KMnO_4$

Write the overall balanced net ionic reaction:

> READ ALL EQUIPMENT
> TO THE CORRECT
> PLACE VALUE AND
> RECORD ALL VALUES
> WITH PROPER UNITS!

▶ ____

	Trial 1	Trial 2	Trial 3
1. Mass of Mohr's Salt			
2. Initial buret reading			
3. Final buret reading			
4. Volume of $KMnO_4$ added			

Calculation of $KMnO_4$ molarity from Trial 1:

Calculation of $KMnO_4$ molarity from Trial 2:

Calculation of $KMnO_4$ molarity from Trial 3 (if necessary):

Average $KMnO_4$ molarity ________________________________

DETERMINATION OF PEROXIDE BY PERMANGANATE

Write the overall balanced net ionic reaction:

▶ ____

	Trial 1	Trial 2
Unknown Number		
1. Mass of empty flask		
2. Mass of flask + sample		
3. Mass of peroxide sample		
4. Initial buret reading		
5. Final buret reading		
6. Volume of $KMnO_4$ delivered		

Trial 1:
    Calculate the grams of H_2O_2 present in the sample.

    Calculate the % H_2O_2 present in the sample.

Trial 2:
    Calculate the grams of H_2O_2 present in the sample.

    Calculate the % H_2O_2 present in the sample.

Average % H_2O_2 in sample ________________________

Name _______________________________

HOMEWORK EXERCISES

Use the net ionic equations in problems 1 and 2 for problems 3-5.

1. a. Write the balanced equations for the half reactions (ionic) and the overall net ionic equation for the titration of Mohr's salt with $KMnO_4$.

Ionic MnO_4^-:

Ionic Fe^{2+}:

Net ionic:

   b. What is the standardized molarity of a $KMnO_4$ solution if 36.31 mL are required to titrate 2.5010 g of Mohr's salt?

2. a. Write the balanced equations for the half reactions and the overall net ionic equation for the titration of H_2O_2 with $KMnO_4$.

Ionic MnO_4^-:

Ionic H_2O_2:

Net ionic:

   b. How is the equivalence point detected in this titration?

3. A 1.6152 g sample of a solution containing H_2O_2 required 11.78 mL of a 0.6134 M $KMnO_4$ solution to reach the equivalence point.
   a. How many grams of H_2O_2 were present in the solution?

   b. What is the percent H_2O_2 present in the solution?

4. 25.5978 g of a 15.5% H_2O_2 solution was titrated with 0.2542 M $Ca(MnO_4)_2$, an oxidizing agent. How many mL of the $Ca(MnO_4)_2$ solution would be required for titration? *Carefully consider the MnO_4^- to $Ca(MnO_4)_2$ mole ratio.*

5. 2.6477 g of the reducing agent $FeSO_4$ was titrated to the equivalence point with 17.25 mL of a $KMnO_4$ solution.
   a. Calculate the molarity of the $KMnO_4$ solution.

   b. What volume (in mL) of the $KMnO_4$ solution would be needed to react with 12.50 grams of 30.0 % H_2O_2 solution?

6. In a redox titration, 12.52 mL of a 0.3264 M $KMnO_4$ solution were required to titrate 1.7832 g of a reducing agent ("X"). If the reducing agent undergoes a 3-electron change per formula unit:
   a. Write the two ionic half reactions and the balanced net ionic equation.

Ionic MnO_4^-:

Ionic Reducing Agent:

Net ionic:

   b. Calculate the molar mass of the reducing agent.

EXPERIMENT 8
HEATS OF REACTION

Calorimetry and the thermochemical study of chemical reactions are introduced. A simple solution calorimeter will be used to determine the heat (enthalpy) of reaction of hydrochloric acid with magnesium metal, the heat of neutralization of hydrochloric acid and sodium hydroxide, the heat of solution of ammonium nitrate in water, and the specific heat of an unknown metal.

Key Chemical Reactions:

$$Mg_{(s)} + 2\,H^{+}_{(aq)} + 2\,Cl^{-}_{(aq)} \rightarrow Mg^{2+}_{(aq)} + 2\,Cl^{-}_{(aq)} + H_{2(g)}$$

$$NaOH_{(aq)} + HCl_{(aq)} \rightarrow NaCl_{(aq)} + H_2O_{(l)}$$

Key Mathematical Equations:

$$q = ms\Delta T = C\Delta T$$

<u>Discussion</u>

<u>ΔH and Heats of Reaction</u>
Most chemical reactions occur with either the liberation or absorption of heat energy. <u>Enthalpy (H) is the value of the heat changes accompanying chemical reactions taking place at constant pressure.</u>

In the study of chemical reactions, the term "heat of reaction" is defined as ΔH or ΔH°_{rxn}, where $\Delta H^{\circ}_{rxn} = \sum \Delta H^{\circ}_{products} - \sum \Delta H^{\circ}_{reactants}$. In an exothermic reaction, heat energy is evolved. This results in a lowering of the enthalpy and a negative numerical value of ΔH. In an endothermic reaction, the value of ΔH is positive and heat must be supplied to complete the reaction.

When physical changes such as changes of state are studied, they are also accompanied by a release or absorption of heat energy. Therefore, these measured heat effects are also referred to as changes in enthalpy, ΔH. For example, the heat of fusion, which is the amount of thermal energy required to melt a solid substance, is represented by the symbol ΔH_{fus}.

<u>Heat Capacity and Specific Heat</u>
The <u>heat capacity</u> (C) of a substance is defined as the amount of heat (usually expressed in calories or joules) required to raise the temperature of a given quantity of the substance by one degree Celsius (or Kelvin). The units are typically J/°C. The heat capacity per unit mass, called <u>specific heat</u> (s), is characteristic of a specific substance. It is the amount of heat required to raise the temperature of one gram of the substance by one degree Celsius. The units are typically J/g-°C.

EXPERIMENT 8   HEATS OF REACTION

Calorimeters

A quantitative measure of the heat released or absorbed accompanying a reaction or physical change can be obtained by containing the reaction in a thermally insulated system of negligible or known heat capacity and measuring the change in temperature of the system. An apparatus in which operations of this type are performed is known as a calorimeter.

A calorimeter should have two basic properties. First, in an ideal calorimeter, the entire heat energy liberated or absorbed in a process is contained, with none leaking out to, or in from, the surroundings. In practice there is no such thing as an ideal calorimeter, but the design of calorimeters is such that the heat losses are small. Second, the calorimeter must have a fixed and precisely known heat capacity so that the heat energy released or absorbed can be determined.

When two substances react (such as in a chemical reaction of a metal and acid or in a neutralization) or when a substance undergoes a physical change, the heat produced or absorbed can be determined by measuring the amounts of the reactants and the change in temperature of the system. The following equations are used:

$$q = ms\Delta T \qquad \text{or} \qquad q = C\Delta T$$

Where $\quad$ $q \quad$ = heat released ($-$) or absorbed ($+$), J
$\qquad\qquad\quad$ $m \quad$ = mass of substance, g
$\qquad\qquad\quad$ $s \quad$ = specific heat of a substance, J/g °C
$\qquad\qquad\quad$ $\Delta T \quad$ = temperature change of the substance, °C
$\qquad\qquad\quad$ $C \quad$ = heat capacity of the calorimeter system, J/°C

Thermochemically, since energy is conserved in a reaction, the heat that is lost by one system is equal to the heat gained by another system, or the total energy of the combined systems must equal zero. For example, if a hot piece of metal is dropped into an insulated calorimeter filled with cold water, the heat lost by the metal must be equal to the heat gained by the water.

If the heat capacity ($C = ms$) of the material of which the calorimeter is constructed can be neglected, only the amount of substances in the calorimeter and the heat involved in the reaction or physical change would determine the temperature change. In theory, the walls and other parts of the calorimeter (such as a thermometer) contribute to the overall enthalpy of reaction. The overall heat evolved or absorbed is the sum of the heat evolved or absorbed for each individual component, $q_{rxn} = \Sigma q$. However, most components in the calorimeter other than the reacting species have low heat capacities and have negligible effects on $\Delta H°_{rxn}$. In this experiment, the heat capacities of the Styrofoam cup, metal stirrer, and thermometer are considered negligible.

The heat of a reaction (q) is related to $\Delta H°_{rxn}$ using the following equation, where n in the number of moles involved in the process:

$$\Delta H^{\circ}_{rxn} = \frac{q}{n}$$

96

EXPERIMENT 8   HEATS OF REACTION

Examples

Example 1:   Heat of Reaction, Heat of Neutralization, Heat of Solution, or Heat of
              Dilution

When 100.0 mL of an HCl solution is mixed with 0.5000 g of Zn powder in a double
Styrofoam cup calorimeter, the temperature of the substances in the calorimeter rose
3.00 °C. Assume the density and specific heat of the resulting solution is that of
water, 1.00 g/mL and 4.184 J/g °C, respectively. The specific heat of solid zinc is
0.393 J/g °C. Calculate the heat of this reaction and the heat of this reaction per mole
of Zn^{2+} formed. The chemical reaction is as follows:

$$Zn_{(s)} + 2\,H^{+}_{(aq)} + 2\,Cl^{-}_{(aq)} \rightarrow Zn^{2+}_{(aq)} + 2\,Cl^{-}_{(aq)} + H_{2\,(g)}$$

The heat generated by this reaction is completely absorbed by the solution, and it is
assumed that no heat is lost to the surroundings.

$$-q_{reaction} = q_{solution}$$

The $q_{solution}$ is calculated from the identities and masses of the substances involved and
the temperature change of the system. Since it can be assumed that no mass is lost in the
reaction, the mass of the reactants must equal the mass of the final solution. The
following equation can then be used:

$$q_{solution} = (m_{Zn2+} + m_{HCl})\,s\,\Delta T$$

$$q_{solution} = (0.5000\ g + 100.0\ g)(4.184\ J/g\ °C)(3.00\ °C) = +1260\ J$$
(heat is absorbed by the solution)

Because $-q_{reaction} = q_{solution}$, $q_{reaction} = -1260$ J (heat is evolved).

It should also be noted that the specific heat of zinc is not needed. When zinc reacts with
HCl, the soluble ion Zn^{2+} is formed, and the mass of the zinc can be combined with the
mass of the water, and the specific heat of water is used.

To calculate the heat of reaction per mole of Zn^{2+} formed, the moles of Zn^{2+} are first
calculated.

$$0.5000\ g\ Zn \times \frac{1\ mole\ Zn}{65.38\ g\ Zn} \times \frac{1\ mole\ Zn^{2+}}{1\ mole\ Zn} = 7.648 \times 10^{-3}\ mole\ Zn^{2+}$$

$$\frac{-1260\ J}{7.648 \times 10^{-3}\ mole\ Zn^{2+}} = -164.7\ kJ/mole\ Zn^{2+}$$

This can be compared to the literature value of -153.9 kJ/mole for $\Delta H^{°}_{rxn,Zn2+}$, and an
experimental error of 7.02 % is calculated. ΔH gives a way to quantitatively compare
two reactions.

97

Example 2:   Heat Changes in Non-Combustion Reactions

A sample of lead (s = 0.130 J/g °C) is heated to 90.00°C and then dropped into a well-insulated Styrofoam cup calorimeter containing 80.00 g of water at 10.00°C. If the final temperature of the mixture is 14.00°C, calculate the mass of the lead sample.

Since energy is conserved, the heat lost by the lead is equal to the heat gained by the water, so that the energy change of the system is 0.

$$- q_{lead} = q_{water}$$

$$- (ms\Delta T)_{lead} = (ms\Delta T)_{water}$$

$$- (\text{mass of lead})(0.130 \text{ J/g °C})(14.00 - 90.00°C) =$$
$$(80.00 \text{ g})(4.184 \text{ J/g °C})(14.00 - 10.00°C)$$

$$\text{mass of lead} = 135 \text{ g}$$

Example 3:   Calculation of Enthalpy Involving Phase Changes
If there is a phase change (for example the melting of ice, which takes place at a constant temperature), the heat of fusion (ΔH_{fus}, the energy required to melt one mole of a solid) of the ice during the phase change must be taken into account. ΔH_{fus} is equal to ml, where the units of l are J/g.

Figure 8.1 shows a typical heating curve for a substance going through two phase changes. For example, ice (s = 2.100 J/g °C) is brought from a temperature below 0 to 0°C, then it undergoes a phase change to water at constant temperature (0°C), then as water (s = 4.184 J/g °C), it is raised from 0°C to 100°C, undergoes a phase change to water vapor (100°C), and then as vapor, raises temperature again.

Temperature vs. Time for a Reaction

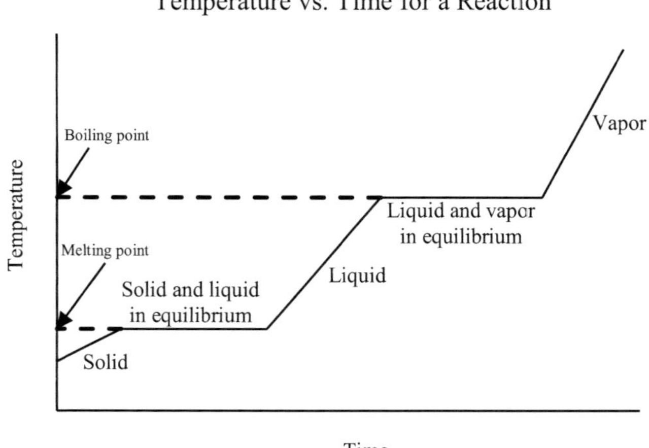

Figure 8.1
Heating Curve

A 0.5000 kg block of ice at −10.0°C is placed in 3.00 kg of water at 20.0°C. At what temperature will the final mixture be? $l_{H_2O} = 333$ J/g.

The heat lost by the water is equal to the heat gained by the ice. The ice temperature will rise from −10.0°C to 0.0°C, then melt to water at constant temperature, and finally the water temperature will be rise from 0.0°C to some final temperature.

$$q_{ice} = -q_{water}$$

$$(ms\Delta T_{-10 \to 0})_{ice} + (ml)_{ice} + (ms\Delta T_{0 \to final\ t})_{ice,\ now\ water} = -(ms\Delta T)_{water}$$

$$(500.0\ g)(2.100\ J/g\ °C)(0.0 - {}^{-}10.0°C) + (500.0\ g)(333\ J/g) +$$
$$(500.0\ g)(4.184\ J/g\ °C)(t_{final} - 0.0°C) = -(3000\ g)(4.184\ J/g\ °C)(t_{final} - 20.0°C)$$

$$T_{final} = 5.10\ °C$$

The mass measurements in this experiment are accurate and easy to obtain. Theoretically, final temperature measurements should be determined graphically. To do this, plot the data (temperature vs. time). Make a straight line through the cooling temperatures and extrapolate back to the initial mixing time line (the y axis). This is shown as the dotted line in Figure 8.2. The temperature value at this intersection is the maximum (or minimum) temperature of the reaction. Figure 8.2 shows this technique.

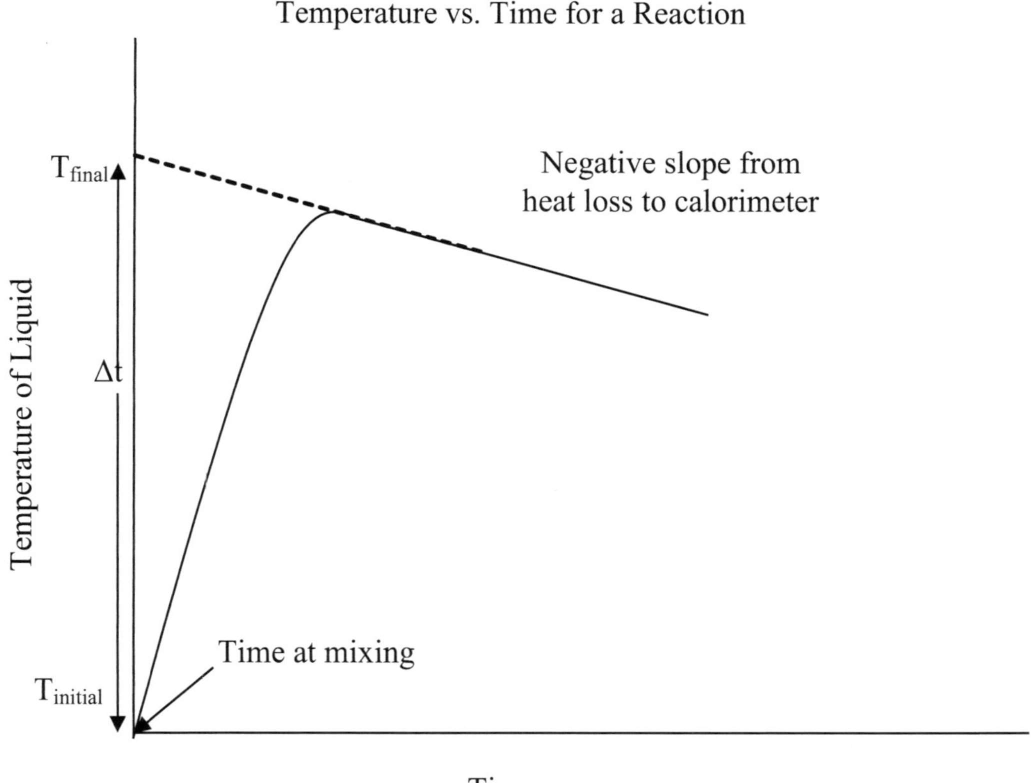

Figure 8.2
Graphical Determination of Maximum Temperature

Although this technique determines the maximum temperature reached, the reactions happen so quickly in this experiment that only the minimum and maximum thermometer readings will be recorded and t_{max} will not be determined graphically.

EXPERIMENT 8   HEATS OF REACTION

<u>Chemicals</u>
    Hydrochloric acid, $HCl_{(aq)}$
    Magnesium metal, $Mg_{(s)}$
    Sodium hydroxide, $NaOH_{(aq)}$
    Ammonium nitrate, $NH_4NO_{3\,(s)}$
    Various unknown metals

<u>Waste</u>
    All waste from this experiment can be disposed down the drain.

<u>Procedure</u>
1. Obtain ONLY 100 mL of the HCl solution in a 150 mL beaker.

2. Normalize two burets: one with the HCl solution and the other with DI water.

3. Assemble the calorimeter as shown in Figure 8.3. Carefully place a thermometer marked in increments of tenths (the range is approximately –5 to 50°C) in the calorimeter through the center hole of the lid. It should be positioned and clamped so that the bulb will be submerged by the liquid, but clear of the bottom of the cup.

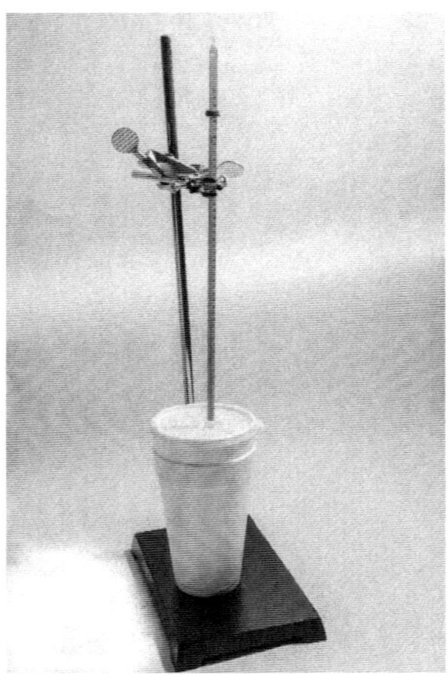

Figure 8.3
Calorimetry Apparatus

4. Using a hot plate, begin heating about 250 mL of water in a 400 mL beaker to boiling (for Part D). Place the thermometer marked in one degree increments (the range is approximately 0 – 100°C) in the beaker. DO NOT USE THE –5 to 50°C THERMOMETER IN THIS STEP: IT WILL BREAK.

5. Weigh the unknown metal sample if the mass is not already noted on the container. Using tweezers or test tube tongs, carefully place the metal in the beaker of water.

6. Continue with the procedure while the water and metal are heating. Add water as necessary to keep the water level around the 250 mL mark.

100

EXPERIMENT 8   HEATS OF REACTION

A. Determination of the Heat of Reaction of Mg and HCl

1. Inspect the calorimeter cup and thermometer to insure they are clean and dry.

2. Deliver approximately 40 mL of the HCl solution to the calorimeter. The amount delivered does not need to be exactly 40 mL, but the amount must be <u>carefully measured and recorded</u>. Reassemble the calorimeter.

3. Obtain a sample of about 0.1000 g (4½ – 5 inches) of Mg ribbon. Sand both sides of the sample so that the ribbon becomes shiny and almost all of the black MgO is removed. Wipe the ribbon with a paper towel after sanding. Fold the ribbon into a bundle and weigh.

4. Allow the HCl temperature to stabilize at least one minute prior to starting the reaction. Record the initial temperature of the HCl.

5. Carefully raise the lid of the calorimeter and insert the bundle of Mg into the calorimeter cup. Quickly reassemble the calorimeter and continue to monitor the temperature, noting that this is an exothermic reaction. Record the maximum temperature reached. Shake or swirl the cup gently during the reaction.

6. Dispose of the solution down the laboratory bench drain. Rinse the calorimeter cup and thermometer and dry thoroughly.

7. Repeat if the percent deviation from the literature value is greater than 20%.

B. Heat of Neutralization of HCl and NaOH

1. Deliver approximately 40 mL of the HCl solution to the <u>dry</u> calorimeter. The amount delivered does not need to be exactly 40 mL, but the amount must be carefully measured and recorded. Reassemble the calorimeter.

2. Normalize a pipet with the standardized NaOH. Carefully pipet 10.00 mL of standardized NaOH to a small beaker. If an autodispenser is used to deliver the NaOH, dispense 10.0 mL directly to the small beaker.

3. Allow the HCl temperature in the calorimeter to stabilize at least one minute prior to starting the reaction and then record the initial temperature of the HCl.

4. Carefully raise the lid of the calorimeter and pour the NaOH into the calorimeter cup. Quickly reassemble the calorimeter and continue to monitor the temperature, noting that this is an exothermic reaction. Record the maximum temperature reached. Shake or swirl the cup gently during the reaction.

5. Dispose of the solution down the laboratory bench drain. Rinse the calorimeter cup and thermometer and dry thoroughly.

6. Repeat if the percent deviation from the literature value is greater than 20%.

EXPERIMENT 8   HEATS OF REACTION

C. *Heat of Solution of Ammonium Nitrate*

1. Deliver approximately 40 mL of DI water to the <u>dry</u> calorimeter. The amount delivered does not need to be exactly 40 mL, but the amount must be carefully measured and recorded. Reassemble the calorimeter.

2. Obtain a sample of about 0.6 g – 0.8 g of NH_4NO_3.

3. Allow the water temperature to stabilize at least one minute prior to starting the reaction and then record the initial temperature of the water.

4. Carefully raise the lid of the calorimeter and transfer the NH_4NO_3 into the calorimeter cup. Quickly reassemble the calorimeter and continue to monitor the temperature, noting that this is an endothermic reaction. Record the minimum temperature reached. Shake or swirl the cup gently during the reaction.

5. Dispose of the solution down the laboratory bench drain. Rinse the calorimeter cup and thermometer and dry thoroughly.

6. Repeat if the percent deviation from the literature value is greater than 20%.

D. *Specific Heat of an Unknown Metal*

1. Deliver approximately 70 mL of DI water to the <u>dry</u> calorimeter. The amount delivered does not need to be exactly 70 mL, but the amount must be carefully measured and recorded. The amount will should be delivered in two increments. Remove the stirrer and reassemble the calorimeter.

2. Allow the water temperature to stabilize at least one minute prior to starting the reaction. Record the initial temperature of the water.

3. Carefully raise the lid of the calorimeter. <u>Quickly</u> take the metal out of the boiling water with tweezers, give the metal a little shake to remove excess water, and place (gently drop) the hot metal into the calorimeter cup to one side of the cup (so it is not touching the thermometer). Quickly reassemble the calorimeter and continue to monitor the temperature with swirling or shaking, noting that this is an exothermic reaction. Record the maximum temperature reached.

4. Dispose of the solution down the laboratory bench drain. Place the metal on a paper towel and dry the metal thoroughly.

5. Return the dry metal to the unknown container.

6. Drain both burets, rinse the one containing HCl with DI water, invert them in the buret clamp, and open the stopcocks.

EXPERIMENT 8   HEATS OF REACTION

LABORATORY REPORT SHEET

Name ______________________________   Date ____________________

Partner ______________________________

READ ALL EQUIPMENT TO
THE CORRECT PLACE VALUE
AND RECORD ALL VALUES
WITH PROPER UNITS!

A.   Heat of Reaction of Mg and HCl

$$Mg_{(s)} + 2\,H^{+}_{(aq)} + 2\,Cl^{-}_{(aq)} \rightarrow Mg^{2+}_{(aq)} + 2\,Cl^{-}_{(aq)} + H_{2(g)}$$

	Value	
Initial HCl solution buret reading		
Final HCl solution buret reading		
Volume of HCl solution added		
Mass of HCl solution added Assume HCl solution density is 1.00 g/mL.		
Specific Heat of HCl solution	4.184 J/g °C	
Mass of Mg		
Initial temperature		
Final temperature		
ΔT		
Calculate the total heat released in Joules by this reaction:		
Total moles of Mg^{2+} formed:		
ΔH (kJ)/mole Mg^{2+} formed:		
Percent deviation from literature value of -466.85 kJ/mole		

B.   Heat of Neutralization between NaOH and HCl

	Value	
Volume of NaOH solution delivered	10.0 mL (if using the autodispenser)	
Mass of NaOH solution Assume NaOH solution density is 1.00 g/mL		
Specific Heat of NaOH solution	4.184 J/g °C	
Molarity of NaOH solution		
Initial HCl solution buret reading		
Final HCl solution buret reading		
Volume of HCl solution added		
Mass of HCl solution added Assume HCl solution density is 1.00 g/mL		
Specific Heat of HCl solution	4.184 J/g °C	
Initial temperature		
Final temperature		
ΔT		
Calculate the total heat released in Joules by this reaction:		
Total moles of NaOH reacted:		
ΔH (kJ)/mole NaOH reacted:		
Percent deviation from literature value of −56.2 kJ/mole		

C.   Heat of Solution of NH_4NO_3 Name________________________________

	Value	
Initial water buret reading		
Final water buret reading		
Volume of water added		
Mass of water Assume density is 1.00 g/mL		
Specific Heat of water	4.184 J/g °C	
Mass of NH_4NO_3		
Molar mass of NH_4NO_3		
Moles of NH_4NO_3		
Initial temperature of water		
Final temperature of water		
ΔT		

Calculate the heat absorbed or lost by the water in J:

Calculate the molar heat of solution in kJ/mol of NH_4NO_3:

Percent deviation from literature value of 25.69 kJ/mol

D.   Specific Heat of Unknown Metal

	Value
Unknown Number	
Initial water buret reading	
Final water buret reading	
Volume of water added	
Mass of water Assume density is 1.00 g/mL	
Specific heat of water	4.184 J/g °C
Mass of metal	
Initial temperature of metal	
Initial temperature of water	
Final temperature	

Metal	s (J/g °C)
Aluminum	0.900
Brass	0.380
Copper	0.385
Iron	0.444
Lead	0.158
Zinc	0.387

Calculate the specific heat of the metal (with units):

Identity of the metal unknown:

Error Analysis

1. In parts A, B, and C, if the calorimeter was not dry, the change in temperature would be too _________ (high/low). This would make the $|\Delta H|$ appear too __________(high/low).

2. If a reaction is exothermic, the sign of ΔH is __________ and heat is ________________.

3. If a reaction is endothermic, the sign of ΔH is __________ and heat is ______________.

4. In the determination of the specific heat of the metal, if some of the water that was used to heat the metal was not removed before the metal was transferred to the calorimeter, the final temperature would be too ________________ (high/low). This would make the specific heat appear too ______________ (high/low). Hint: put in hypothetical numbers.

EXPERIMENT 8   HEATS OF REACTION

Name _______________________

<u>HOMEWORK EXERCISES</u>

This homework is three pages

1. Magnesium solid reacts with aqueous hydrochloric acid to form the Mg^{2+} ion in solution. In an experiment, 60.0 mL of aqueous HCl was mixed with 0.1297 g of magnesium solid in a double Styrofoam cup calorimeter. The reaction caused the temperature of the substances in the calorimeter to rise 10.02°C. Assume the density and specific heat of the HCl solution is that of water, 1.00 g/mL and 4.184 J/g °C, respectively. The specific heat of magnesium is 1.02 J/g °C.

    a. Write the balanced chemical equation for this reaction.

    b. Calculate the heat of this reaction, ΔH_{rxn}, in kJ.

    c. Calculate the heat of this reaction per mole of Mg^{2+} formed, ΔH_{rxn}/mole Mg^{2+}.

    d. The literature value of this reaction is –466.85 kJ/mole. Calculate the percent deviation of the experimental value from the literature value.

2. A 45.5 kg copper rod (s = 0.385 J/g °C) is heated to 475.0°C and then quenched into a well-insulated water bath containing 1000.0 kg of water at 25.0°C. What is the final temperature of the mixture?

107

3. A student goes to the cafeteria to get an iced tea. He puts 3 ice cubes (approximately 20.0 mL each, s_{ice} = 2.100 J/g °C) from the ice machine (at − 15.0°C) into a well-insulated cup and then fills the cup with 375 mL of tea. The tea is coming from a machine at room temperature, which is 22.5°C. The student is so immersed in studying chemistry that he forgets about his drink. When he finally goes to take his first drink, he notices that the ice has just melted. At what temperature is his iced tea? Assume the density of ice is that of water. It may be helpful to plot the temperatures on the heating curve below.

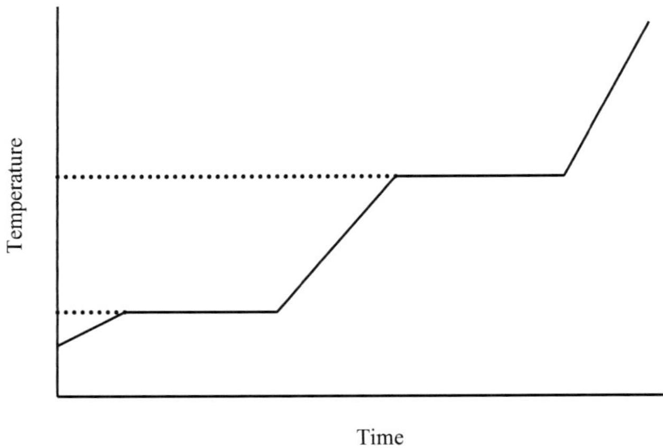

4. In a neutralization reaction in a well-insulated calorimeter, 112.9 g of 1.2915 M NaOH were added to 129.1 g of 1.1294 M HCl. The following data were obtained. Assume the density of each solution is 1.00 g/mL.

Time (s)	Temperature (°C)
0	20.00
5	24.00
10	27.50
15	27.20
20	26.90
25	26.60
30	26.20
35	25.90
40	25.60
45	25.30
50	25.00
55	24.60
60	24.30

a. Using the graphing technique outlined in the discussion (see Figure 8.2), determine the maximum temperature graphically <u>on graph paper</u>.
Clearly mark the maximum temperature on the graph.

b. Assuming the calorimeter, thermometer, and stirrer have a negligible impact on the ΔH_{rxn}, calculate the ΔH_{rxn} (in kJ) for this reaction.

c. Calculate the ΔH_{rxn} per mole of NaOH reacted.

d. If the Styrofoam calorimeter (s = 1.131 J/g °C) weighs 30.0 g, the glass thermometer (s = 0.840 J/g °C) weighs 20.0 g, and the aluminum stirrer (s = 0.900 J/g °C) weighs 10.0 g, calculate the ΔH_{rxn} per mole of NaOH reacted.

e. The literature value of this reaction is –56.2 kJ/mole. <u>Qualitatively</u> compare the ΔH_{rxn} from (c) and (d) and comment on the effects the calorimeter, thermometer, and stirrer have on the ΔH_{rxn} and if they can be considered negligible in the calculations.

EXPERIMENT 9
GAS LAWS

The ideal gas law equation will be demonstrated with two simple experiments: the decomposition of hydrogen peroxide and the reaction of aluminum and hydrochloric acid. The application of Dalton's law of partial pressures to the pressure term in the ideal gas equation will be investigated.

Key Chemical Reactions:

$$2\ Al_{(s)} + 6\ HCl_{(aq)} \rightarrow 2\ AlCl_{3\ (aq)} + 3\ H_{2\ (g)}$$

$$2\ H_2O_{2\ (aq)} \rightarrow 2\ H_2O_{(l)} + O_{2\ (g)}$$

Key Mathematical Equations:

$$PV = nRT$$

Dalton's Law of Partial Pressures: $P_{total} = P_{gas} + P_{H_2O}$

Discussion

The pressure, volume, amount, and temperature of a gas are related in the ideal gas law, given by the equation:

$$PV = nRT$$

Where

P is the pressure of the gas
V is the volume of the gas
n is the moles of gas, and
R is the universal gas constant

The units of all variables must be converted to match the appropriate value of R. Frequently used R values are given below:

$$0.0821\ \frac{L\ atm}{mole\ K} \qquad 62.4\ \frac{L\ torr}{mole\ K} \qquad 8.314\ \frac{J}{mole\ K} \qquad 8.314\ \frac{m^3\ Pa}{mole\ K}$$

The ideal gas law can be rearranged to determine the density or molar mass of the gas with these terms:

$$\text{molar mass of gas} = \frac{\text{mass of gas}}{\text{moles of gas}} \qquad \text{and} \qquad \text{density } (\rho) \text{ of the gas} = \frac{\text{mass of the gas}}{\text{volume of the gas}}$$

Substituting these values into the ideal gas equation gives other ways to present the ideal gas law:

$$PV = \frac{gRT}{MM} \qquad \text{and} \qquad P = \frac{\rho RT}{MM}$$

When gases are investigated in an experiment, they are often collected by the displacement of a liquid in a container. For example, in the reaction of zinc with sulfuric acid, hydrogen gas is generated, and the gas is collected by displacing water in a collection bottle.   The gas in the collection vessel is a mixture of hydrogen and water vapor, and the total pressure in the collection vessel is the combined partial pressures of hydrogen and water.   Since only the pressure of hydrogen is needed, Dalton's law of partial pressures must be used to calculate the pressure exerted only by the hydrogen. Therefore, the pressure due to the water vapor must be subtracted from the total (atmospheric) pressure, the pressure read at the barometer.   The vapor pressure of water can be read or interpolated from values found in the Appendix.

$$P_{Total\ (atmospheric)} = P_{H_2} + P_{H_2O}$$

Example:

$$Zn_{(s)} + H_2SO_{4\ (aq)} \rightarrow ZnSO_{4\ (aq)} + H_{2\ (g)}$$

2.00 g of zinc are reacted with sulfuric acid as shown in the chemical reaction presented above.   The gas generated from this reaction is collected over water at 27.0 °C.   The pressure at the barometer is recorded as 742.6 mm Hg.   Calculate the theoretical volume of the gas collected.

Since the ideal gas equation can only be used for gases, the theoretical moles of hydrogen must first be calculated.

$$2.00 \text{ g Zn} \times \frac{1 \text{ mol Zn}}{65.39 \text{ g Zn}} \times \frac{1 \text{ mol H}_2}{1 \text{ mol Zn}} = 0.0306 \text{ mol H}_2$$

Since the hydrogen is collected over water, Dalton's law must be used to calculate the pressure of hydrogen.   The vapor pressure of water at 27.0 °C can be found in the Appendix.

$$P_{Total\ (atmospheric)} = P_{H_2} + P_{H_2O}$$

$$742.6 \text{ mm Hg} = P_{H_2} + 26.7 \text{ mm Hg}$$

$$P_{H_2} = 715.9 \text{ mm Hg}$$

$$V = \frac{nRT}{P} = \frac{(0.0306 \text{ mol H}_2)(0.0821 \frac{L\ atm}{mol\ K})(300.0 \text{ K})}{(715.9 \text{ mm Hg} \times \frac{1 \text{ atm}}{760 \text{ mm Hg}})} = 0.800 \text{ L}$$

<u>Chemicals</u>

Aluminum	$Al_{(s)}$
Hydrochloric acid (6 M)	$HCl_{(aq)}$
Hydrogen peroxide	$H_2O_{2\ (aq)}$
Yeast	
Oxygen	$O_{2\ (g)}$
Hydrogen	$H_{2\ (g)}$

EXPERIMENT 9   GAS LAWS

<u>Waste</u>

All waste can be disposed down the drain.

<u>Procedure</u>

Refer to the apparatus set up as shown in Figure 9.1 for the equipment set up.

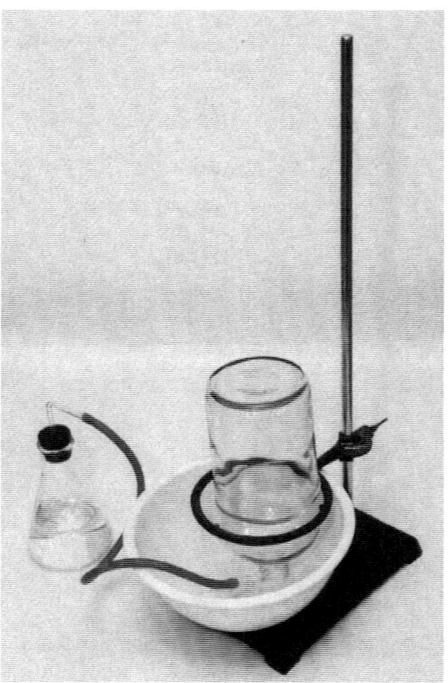

Figure 9.1
Gas Laws Set Up

<u>Experiment 1: Decomposition of Hydrogen Peroxide</u>

1. Turn the ring stand around and place a ring clamp on it so the ring is facing the front of the bench. Fill the plastic bowl to 1 inch from the rim with tap water and place it under the ring clamp. Place a rubber band around the body of a 1 L glass collection bottle. Fill the bottle with tap water and then use the DI water bottle to completely fill the bottle so that a dome of water is present on the mouth of the bottle. Cover the mouth of the bottle temporarily with the wide end of a large rubber stopper and invert the bottle and stopper into the bowl so the neck of the bottle is under water. Remove the stopper when done. While one partner is holding the bottle, the other partner should lower the large ring clamp over the body of the bottle and tighten the ring clamp to the ring stand at the rim of the bowl. It may be helpful to insert some wet paper towels between the bottle and the ring so that the bottle does not slip in the bowl. No air bubbles (or a very tiny one) should be present in the bottle when inverted. Insert the glass end of the gas delivery tubing into the mouth of the bottle.

2. Set up the aspirator tubing to remove displaced water, placing the free end in the bowl.

3. Add exactly 50.0 mL of DI water from a graduated cylinder to a 250 mL Erlenmeyer flask.

4. Add 3 small spatulafuls of yeast to the 250 mL Erlenmeyer flask. Connect the flask to the system by placing the stopper with the gas delivery tubing on the top of the flask.

5. Bring a *dry* 50 mL Erlenmeyer flask and a #1 stopper to the autodispenser. Dispense 20.0 mL of the unknown hydrogen peroxide solution to the 50 mL Erlenmeyer flask, and promptly stopper the 50 mL flask. Record the volume on Data Sheet A.

6. Lift the stopper of the 250 mL Erlenmeyer flask and pour in the hydrogen peroxide quickly, promptly replacing the stopper on the 250 mL flask. The reaction will start immediately.

7. Allow the reaction to proceed, occasionally swirling the flask. Use the aspirator as necessary to prevent the displaced water from overflowing the bowl. Qualitatively note the temperature of the contents of the flask.

8. After the gas generation has ceased, adjust the water level in the bottle to the same height as the water in the bowl by either raising or lowering the collection bottle or adding more water to the plastic bowl. When the water levels inside and outside the collection bottle are equal, mark the level on the bottle by adjusting the rubber band.

9. Remove the bottle and allow the water to drain out, taking care that the marker is not disturbed. Fill the bottle to the marker with tap water and then transfer the water to the large graduated cylinder to measure the volume. This represents the volume of gas collected.

10. Record the atmospheric pressure and measure the temperature of the water in the bowl. This temperature will be assumed to be the same as that of the gas collected over the water.

11. Set up and start Experiment 2 (steps 1-8) *before starting Data Sheet A's calculations and questions*. The calculations and questions in Data Sheet A can be completed while waiting for the aluminum and hydrochloric acid reaction to start.

Experiment 2: Aluminum and Hydrogen Chloride

1. Set up the experiment again, as described in step 1 of Experiment 1, above. It is not necessary to obtain new water in the bowl; just add to it if necessary.

2. Set up the aspirator tubing to remove displaced water, placing the free end in the bowl.

3. Weigh approximately 0.5 grams of aluminum (~4" x 4"). Record the mass on Data Sheet B.

4. Add exactly 50.0 mL of DI water from a graduated cylinder to a 250 mL Erlenmeyer flask.

5. Carefully measure 25 mL of 6 M HCl and add to the flask. Connect the flask to the system by placing the stopper with the gas delivery tubing on the top of the flask.

6. Tear the aluminum into four smaller pieces and crumple into loose balls.

7. Lift the stopper and add the four aluminum balls and then replace the stopper.

8. Allow the reaction to proceed, occasionally swirling the flask. The reaction of aluminum and hydrochloric acid will start to be visible between 10 and 15 minutes, and the entire reaction will take between 20 and 30 minutes. Complete the calculations and questions for the first experiment during this time. Use the aspirator as necessary to prevent the displaced water from overflowing the bowl.

9. After the gas generation has ceased, adjust the water level in the bottle to the same height as the water in the bowl by either raising or lowering the collection bottle or adding more water to the plastic bowl. When the water levels inside and outside the collection bottle are equal, mark the level on the bottle by adjusting the rubber band.

10. Remove the bottle and allow the water to drain out, taking care that the marker is not disturbed. Fill the bottle to the marker with tap water and then transfer the water to the large graduated cylinder to measure the volume. This represents the volume of gas collected.

11. Record the atmospheric pressure and measure the temperature of the water in the bowl. This temperature will be assumed to be the same as that of the gas collected over the water.

12. Rinse the graduated cylinder (HCl was last in it) several times with tap water.

13. Complete Data Sheet B and associated questions, as well as the Error Analysis exercise on the last page of this lab.

EXPERIMENT 9   GAS LAWS

DATA SHEET A

Name______________________________________   Date________________

R value used: ______________________________

**READ ALL EQUIPMENT TO THE
CORRECT PLACE VALUE AND RECORD
ALL VALUES WITH PROPER UNITS!**

Write the balanced chemical equation of the decomposition of hydrogen peroxide into water and oxygen (include states):

➢ ____

Unknown #: __________   Volume of unknown H_2O_2 dispensed: ______________________

Mass of unknown H_2O_2 solution dispensed (density = 1.00 g/mL): ___________________

	Value (with units)	Value (with units matching R value)
Volume of wet oxygen collected		
Temperature		
Atmospheric pressure		
Partial pressure of water at experiment temperature (record/interpolate this value in <u>mm Hg</u> from the Appendix) *Show interpolation calculation:*		
Partial pressure of oxygen at experiment temperature		

Moles of oxygen collected:

Mass of H_2O_2 in sample:

% H_2O_2 in unknown solution:

117

Questions:

1. Based on the moles of oxygen collected, calculate the grams of oxygen collected.

2. Calculate the density of oxygen gas in kg/m^3.

3. Using the ideal gas law, calculate the moles of oxygen collected if Dalton's law was not used.

Not using Dalton's law would make the % H_2O_2 in the unknown appear ______________.

greater/smaller

4. The solubility of oxygen in water is given in the following diagram.
    a. Give the graph a title.

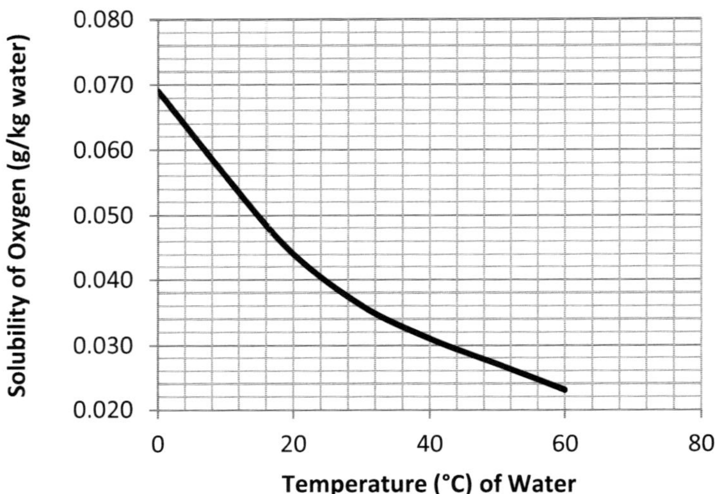

    b. Using the above graph, determine the total mass of oxygen dissolved in the solution in the reaction flask for this experiment. (This is the amount of O_2 that did not get collected). Assume that the solution (H_2O_2 and water) is all water when calculating the total volume of water in the flask, and the density of the solution is 1.00 g/mL.

    c. Calculate the % O_2 dissolved. (Total O_2 here includes both collected and dissolved.)

5. Is this reaction exothermic or endothermic? Briefly explain.

EXPERIMENT 9   GAS LAWS

DATA SHEET B

Name____ Date______________

R value used: ____________________________

**READ ALL EQUIPMENT TO THE
CORRECT PLACE VALUE AND RECORD
ALL VALUES WITH PROPER UNITS!**

Write the balanced chemical equation of aluminum and hydrochloric acid (include states):

➢ ____

Mass of aluminum reacted: ____________________________

Theoretical moles of hydrogen generated based on the mass of aluminum:

	Value (with units)	Value (with units matching R value)
Temperature		
Atmospheric pressure		
Partial pressure of water at experiment temperature (record/interpolate this value in <u>mm Hg</u> from the Appendix) *Show interpolation calculation:*		
Partial pressure of hydrogen at experiment temperature		

Theoretical volume of hydrogen generated:

Experimental volume of hydrogen collected: ____________________________

% Error:

Questions:
1. Based on the theoretical moles of hydrogen, calculate the grams of hydrogen generated.

2. Using the theoretical volume of hydrogen, calculate the density of hydrogen gas in kg/m^3.

3. Using the ideal gas law, calculate the theoretical volume of hydrogen if Dalton's law was not used.

    Not using Dalton's law will make the theoretical volume appear ______________.
    greater/smaller

4. The solubility of hydrogen in water is given in the following diagram.
    a. Give the graph a title.

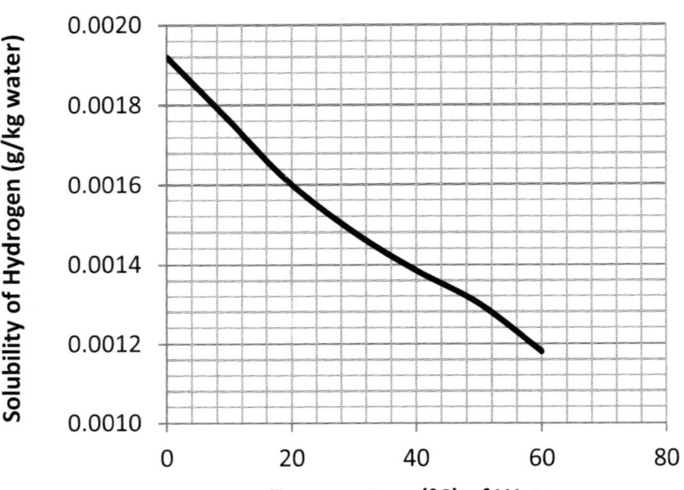

    b. Using the above graph, determine the total mass of hydrogen dissolved in the solution in the reaction flask for this experiment. (This is the mass of hydrogen that did not get collected).   Assume that the solution (HCl and water) is all water when calculating the total volume of water in the flask, and the density of the solution is 1.00 g/mL.

    c. Calculate the % hydrogen dissolved. (The total H_2 here is only collected gas.)

Error Analysis

Use the following balanced chemical reaction for this question:

$$CaCO_3 \, (s) + H_2SO_4 \, (aq) \rightarrow H_2O \, (l) + CaSO_4 \, (s) + CO_2 \, (g)$$

1. Graph the following data of the amount of CO_2 dissolved in water:

T (°C)	g CO_2/kg water
0	3.30
10	2.50
20	1.70
30	1.25
40	1.00
50	0.75
60	0.60

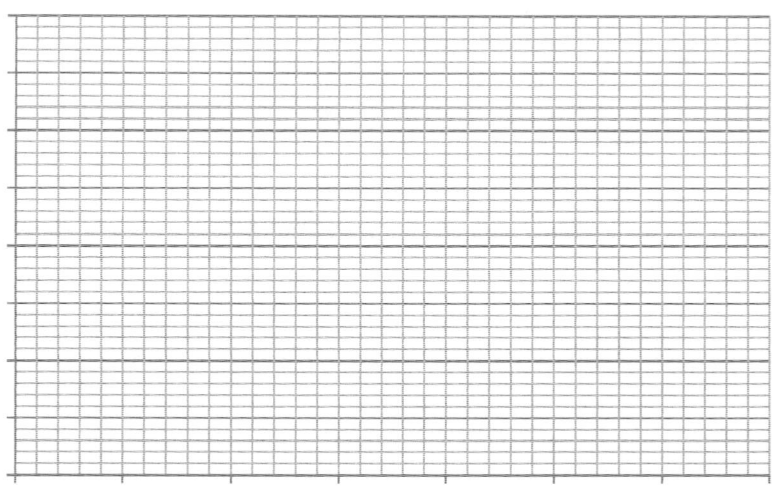

2. 0.5000 g of $CaCO_3$ (the limiting reagent) react with 100.0 mL of 0.0800 M sulfuric acid at 751.8 mm Hg and 23.6 °C. The gas is collected over water.

   a. Calculate the mass of CO_2 produced from the reaction of 0.5000 g of $CaCO_3$.

   b. Using the graph, determine the mass of CO_2 dissolved in solution. Assume the density of the solution is 1.00 g/mL and that this solution is all water.

   c. Determine the mass of CO_2 collected (this is the difference between produced and dissolved CO_2).

   d. Using the ideal gas equation, determine the volume of CO_2 collected.

Name ________________________________

HOMEWORK EXERCISES

1. Write the balanced chemical equation of the decomposition of hydrogen peroxide into water and oxygen (include states):

2. Write the balanced chemical equation of the reaction of aluminum and hydrochloric acid, which produces a salt and hydrogen gas (include states):

3. a. Magnesium solid reacts with 6 M hydrochloric acid to form magnesium chloride and hydrogen. Write the balanced chemical equation, with states, for this reaction.

    b. What volume of hydrogen gas, collected over water at a barometric pressure of 29.67 inches Hg and a temperature of 21.4 °C, would be produced by the reaction of 0.9185 g of magnesium?

Continued

4. a. Write the balanced equation, with states, for the reaction of sodium sulfite with hydrochloric acid. Refer to the acid reactions that produce gas in the Solubility Rules experiment of this manual.

   b. From the following data and known identity of the gas, calculate the experimental value of the density of the gas formed. The gas is collected by the displacement of water at 23.0 °C, and the barometer reading is 0.9822 atm. The volume of the gas collected is 316 mL.

   c. Calculate the percent error of this experiment from the published density of 2.63 g/L.

   d. Convert the experimental value of the density to g/cm^3.

EXPERIMENT 10
BONDING AND MOLECULAR GEOMETRY

The way the atoms in a compound are bonded together helps determine the shape, or molecular geometry, of the compound. Knowing the three dimensional arrangement of atoms assists in predicting the physical and chemical properties of the compound.

To determine how the atoms are bonded together, possible Lewis structures are formulated. Formal charges on these Lewis structures can be calculated to determine the most probable arrangement of the atoms within a molecule. Once the correct Lewis Structure has been determined, VSEPR (Valence Shell Electron Pair Repulsion) theory is used to predict the shape of the molecule as well as the bond angles between the atoms. Knowing Lewis structures for covalently bonded compounds and polyatomic ions is an important tool to determine the compound's polarity, chemical reactivity, hybridization, and resonance. The concept of hybridization can then be used to show how the covalent bonds within the compound are formed. The following flow chart describes this process.

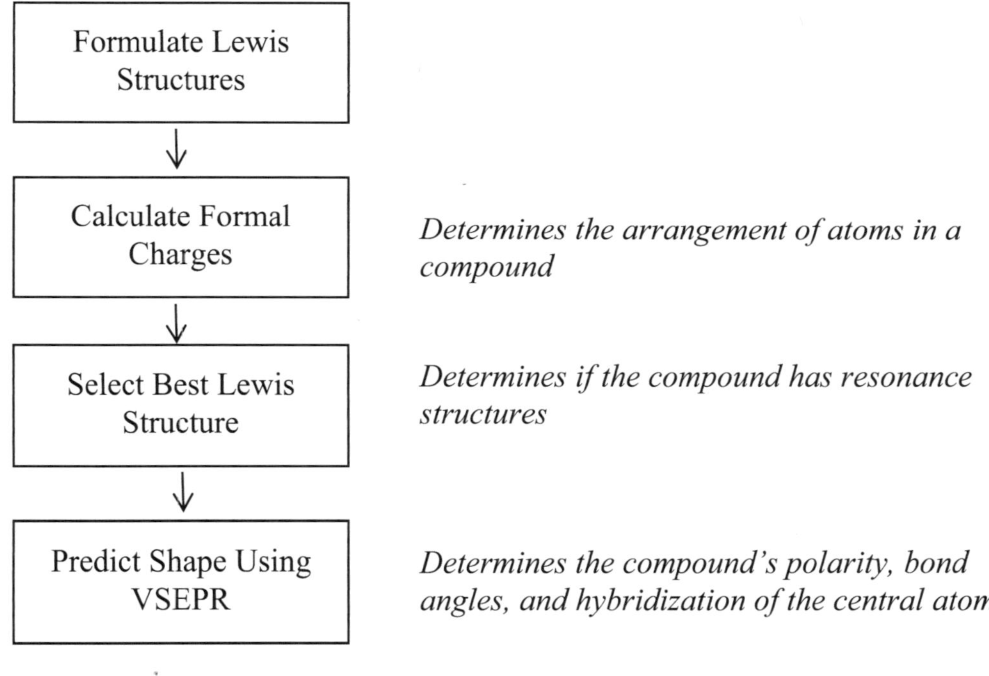

Key Chemical Reactions: None

Key Mathematical Equations: None

Fundamental Chemistry Concepts:
        Arrangement of Electron Pairs and Molecular Geometry
        Mastery of Table 10.1 is required

Discussion
Lewis Structures
A Lewis dot symbol is the symbol of an element with one dot for each valence electron (outer shell or highest principle quantum number) of the element. The number of valence electrons for representative elements is equal to the group number of the element. For example, nitrogen in group 5A has five valence electrons, and is represented by the following symbol.

$$\cdot \ddot{N} \cdot$$

Likewise, noble gases (group 8A) have eight electrons in their outer shell, and are very stable. For most compounds, maximum stability of an atom within a compound occurs when the atom is isoelectronic (has the same number of electrons) as a noble gas.

Lewis structures simply describe how atoms are bonded to each other in a molecule. In 1916, Gilbert N. Lewis proposed that atoms form bonds by sharing pairs of electrons between them. For ionic compounds, the electrons are completely transferred to one of the elements so that both atoms are isoelectronic with noble gases. In the following example on the reactant side of the equation, the atoms show the valence electrons; on the product side, the chlorine is shown with all the electrons in brackets, signifying a complete electron transfer of sodium's one electron. Note the charges on the products are due to this electron transfer.

$$Na\cdot \;\; + \;\; \cdot\ddot{\underset{..}{Cl}}: \;\; \rightarrow \;\; Na^{+} \;\; + \;\; \left[:\ddot{\underset{..}{Cl}}: \right]^{-}$$

For covalent compounds, the electrons are shared by both of the atoms. Many atoms obey the octet rule which states that when atoms form covalent bonds, they share electrons to achieve an outer shell that has eight electrons. A covalently bonded compound is shown as follows:

$$:\ddot{\underset{..}{Cl}}\cdot \;\; + \;\; \cdot\ddot{\underset{..}{Cl}}: \;\; \rightarrow \;\; :\ddot{\underset{..}{Cl}}:\ddot{\underset{..}{Cl}}:$$

In a single covalent bond, two elements share two electrons (symbol: one line). In a double bond, two elements share four electrons (symbol: two lines). A triple bond is formed when two elements share six electrons (symbol: three lines). The Cl_2 molecule drawn above is represented by the following structure. Oftentimes the outer electrons are not shown.

$$Cl \;\; + \;\; Cl \;\; \rightarrow \;\; Cl\text{-}Cl$$

Electrons are classified as either nonbonding or bonding and are almost always grouped in pairs. Bonding electrons are shared by two atoms. Nonbonding electrons are not involved in bonding but are part of the atom's outer shell of electrons. Paired nonbonding electrons are known as lone pairs. Most elements obey the octet rule when bonding by having eight electrons (including shared electrons) in their outer shell.

A few atoms are exceptions to the octet rule:
1. Hydrogen requires only two electrons (one bond) in its outer shell.
2. Group I elements require only two electrons in their outer shell; Group II elements require only four electrons in their outer shell; and Group III elements require only six electrons in their outer shell. Other elements, such as tin, can also be an exception to the octet rule.
3. Atoms in Row 3 and below may have expanded octets where there are more than eight electrons in their outer shell. Examples include phosphorus, sulfur, xenon, and chlorine.
4. Metals in groups 1A, 2A, 3A, 4A, and 2B have 1, 2, 3, 4, and 2 electrons in their outer shell, respectively, and do not need to have a full octet.

Drawing Lewis structures can be done easily using the following method:
1. Draw a skeletal structure (arrangement of atoms) of the molecule. Start by placing the least electronegative element as the central atom. Hydrogen, although it may be the least electronegative element in a compound, is never a central element.

    Electronegativity is the attraction of an atom for the electrons in a bond. Linus Pauling (1901 – 1994) developed relative numerical values for each element's electronegativity. Electronegativity increases within a group from bottom to top and within a period from left to right. Electronegativity values are relative; there are a number of different scales.

2. Count all valence electrons. (Refer to the periodic table for this – in the representative elements, the number of valence electrons is equal to the group in which the element belongs.) If the molecule is a polyatomic ion, add an electron for each negative charge or subtract an electron for each positive charge.

3. Place two electrons in each bond.

4. Continue to distribute remaining electrons by completing the octets of the atoms attached to the central atom, adding electrons in pairs. Refer to the above octet rule exceptions for elements that may not have exactly eight electrons in their outer shell.

5. Place any remaining electrons on the central atom in pairs.

6. If the central atom does not have an octet but requires one, form double or triple bonds by moving an outer atom's electrons to the bond with the central atom.

For example, drawing carbon tetrachloride, CCl_4,

$$Cl$$

Cl   C   Cl   is drawn and the total number of valence electrons
                    is calculated. (4 Cl atoms x 7 e⁻ per Cl atom +
Cl   1 C atom x 4 e⁻ per C atom = 32 electrons)

Two electrons are placed in each of the bonds with carbon and then the octets of the outer atoms are completed giving

$$\ddot{Cl}$$

$$:\ddot{Cl}: \quad C \quad :\ddot{Cl}: \qquad \text{or more simply,} \qquad Cl-C-Cl$$

$$:\ddot{Cl}:$$

All valence electrons are used and each atom has a complete octet of electrons.

For CO_2,   O   C   O   is drawn and the total number of valence
                                  electrons is calculated. (2 O atoms x 6 e⁻
                                  per O atom + 1 C atom x 4 e⁻ per C atom
                                  = 16 electrons)

Two electrons are placed in each of the bonds with carbon and then the octets of the outer atoms are completed giving

$$:\ddot{O}: \quad C \quad :\ddot{O}:$$

This structure uses all of the electrons, but carbon does not have a complete octet. Therefore, a double bond is made with each of the oxygen atoms by moving two electrons from an outer oxygen atom to the carbon-oxygen bond as such:

$$:\ddot{O}::C::\ddot{O}:$$

Formal Charge
The formal charge of an atom is the apparent charge on the atom in a covalent bond. It is the electrical charge difference between the valence electrons of the isolated atom and the number of electrons assigned to that atom in a Lewis structure. Determining the formal charge on an atom helps select the most plausible Lewis structure of the molecule. Molecules with small or no formal charges on the atoms are more likely to exist than molecule with large or many formal charges. The formal charge on an atom can be determined by the following equation:

$$\text{Formal charge on atom} = \text{Number of valence } e^- \text{ in the atom} - \tfrac{1}{2}\left\{\text{Number of bonding } e^- \text{ of atom}\right\} - \left\{\text{Number of non-bonding } e^- \text{ of atom}\right\}$$

The number of bonding electrons are the total number of electrons shared in a covalent bond between elements. A single bond has two electrons in the bond, a double bond has four electrons, and a triple bond has six electrons. The number of non-bonding electrons is the number of electrons not involved in bonding, typically the lone pairs.

For example, consider nitric acid, HNO_3. Since the least electronegative atom (excluding hydrogen) in this molecule is N, it is the central atom. Two structures are possible:

(A)   (B)

The formal charges are calculated as follows:

<table>
<tr><td colspan="2" align="center"><u>Structure A</u></td><td colspan="2" align="center"><u>Structure B</u></td></tr>
<tr><td colspan="2"><u>N</u></td><td colspan="2"><u>N</u></td></tr>
<tr><td>Valence e^-</td><td>5</td><td>Valence e^-</td><td>5</td></tr>
<tr><td>$-\tfrac{1}{2}$ bonding e^-</td><td>4</td><td>$-\tfrac{1}{2}$ bonding e^-</td><td>4</td></tr>
<tr><td>$-$ non bonding e^-</td><td>0</td><td>$-$ non bonding e^-</td><td>0</td></tr>
<tr><td>Formal Charge</td><td>+1</td><td>Formal Charge</td><td>+1</td></tr>
<tr><td colspan="2"><u>O (top, attached to N)</u></td><td colspan="2"><u>O (double bonded)</u></td></tr>
<tr><td>Valence e^-</td><td>6</td><td>Valence e^-</td><td>6</td></tr>
<tr><td>$-\tfrac{1}{2}$ bonding e^-</td><td>1</td><td>$-\tfrac{1}{2}$ bonding e^-</td><td>2</td></tr>
<tr><td>$-$ non bonding e^-</td><td>6</td><td>$-$ non bonding e^-</td><td>4</td></tr>
<tr><td>Formal Charge</td><td>−1</td><td>Formal Charge</td><td>0</td></tr>
<tr><td colspan="2"><u>O (attached to H)</u></td><td colspan="2"><u>O (attached to H)</u></td></tr>
<tr><td>Valence e^-</td><td>6</td><td>Valence e^-</td><td>6</td></tr>
<tr><td>$-\tfrac{1}{2}$ bonding e^-</td><td>3</td><td>$-\tfrac{1}{2}$ bonding e^-</td><td>2</td></tr>
<tr><td>$-$ non bonding e^-</td><td>2</td><td>$-$ non bonding e^-</td><td>4</td></tr>
<tr><td>Formal Charge</td><td>+1</td><td>Formal Charge</td><td>0</td></tr>
<tr><td colspan="2"><u>O (right, attached to N)</u></td><td colspan="2"><u>O (single bonded, not attached to H)</u></td></tr>
<tr><td>Valence e^-</td><td>6</td><td>Valence e^-</td><td>6</td></tr>
<tr><td>$-\tfrac{1}{2}$ bonding e^-</td><td>1</td><td>$-\tfrac{1}{2}$ bonding e^-</td><td>1</td></tr>
<tr><td>$-$ non bonding e^-</td><td>6</td><td>$-$ non bonding e^-</td><td>6</td></tr>
<tr><td>Formal Charge</td><td>−1</td><td>Formal Charge</td><td>−1</td></tr>
<tr><td colspan="2"><u>H</u></td><td colspan="2"><u>H</u></td></tr>
<tr><td>Valence e^-</td><td>1</td><td>Valence e^-</td><td>1</td></tr>
<tr><td>$-\tfrac{1}{2}$ bonding e^-</td><td>1</td><td>$-\tfrac{1}{2}$ bonding e^-</td><td>1</td></tr>
<tr><td>$-$ non bonding e^-</td><td>0</td><td>$-$ non bonding e^-</td><td>0</td></tr>
<tr><td>Formal Charge</td><td>0</td><td>Formal Charge</td><td>0</td></tr>
</table>

Formal charges are written next to each atom as follows:

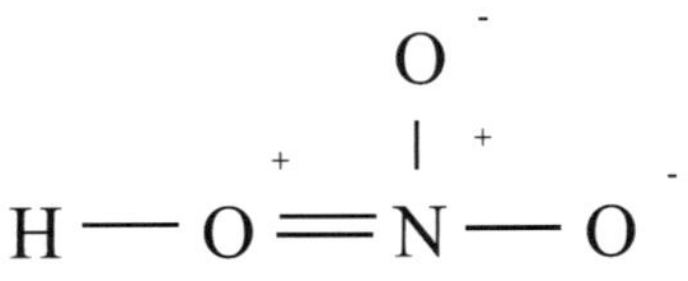

$$\overset{\cdot}{\underset{(A)}{H-\overset{+}{O}=\overset{|}{\overset{+}{N}}-\overset{\cdot}{O}}}\qquad\qquad\overset{\cdot}{\underset{(B)}{H-O-\overset{|}{\overset{+}{N}}=O}}$$

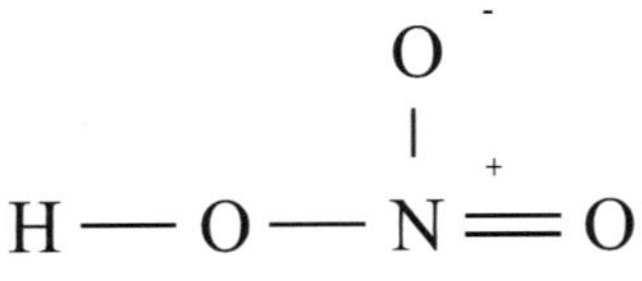

It can be seen that structure (B) has fewer formal charges than structure (A) (2 vs. 4). Therefore, Structure B is the preferred structure. Additionally, the sum of the formal charges on the correct structure must equal the net charge on the molecule. In structure B, the sum of the formal charges is 0 (+1 and −1), which is the overall charge of HNO_3. Structure A also gives a 0 net charge, but there are more formal charges on A, so it is not the preferred structure.

Two other rules are helpful in determining the most plausible structure of a molecule:
1. If there appears to be two structures of a molecule with the same number and same values of formal charges, the most viable structure is the one in which the negative formal charges are placed on the most electronegative elements.
2. There are two distinct schools of thought when it comes to determining a plausible structure for molecules that can have expanded octets. Either of the following could be given when determining a Lewis structure, but the specific reason must be stated. Both structures will give the same molecular geometry.
   (a)   The arrangement satisfies the octet rule, even though it may have greater or more formal charges, or
   (b)   The arrangement minimizes the formal charge, although an expanded octet may be present.

It should also be noted that formal charges are not charge separations in the molecule, but they merely assist in keeping track of the valence electrons and help determine the preferred structure.

Resonance
When a double or triple bond is present, one Lewis structure may not accurately represent the molecule, and two or more structures may be drawn to represent the molecule. These multiple structures are termed resonance structures. This is a human invention that describes a real molecule in terms of two or more nonexistent molecules. Structures that have resonance do not oscillate between structures, but there is a unique structure that is an average between them.

For example, the sulfite ion, SO_3^{2-}, can be shown as any of the three given structures:

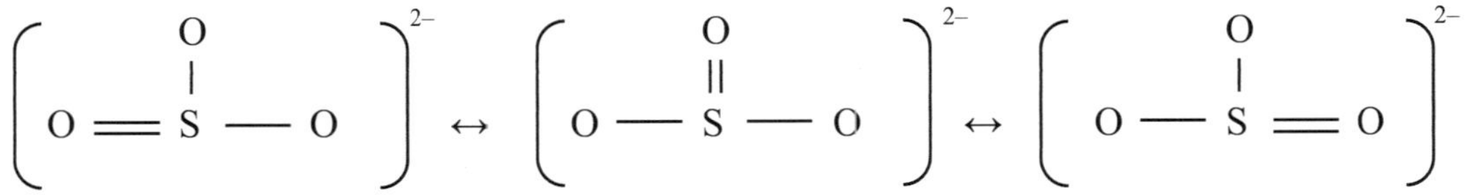

$$\left[\overset{O}{\underset{O=S-O}{|}}\right]^{2-}\leftrightarrow\left[\overset{O}{\underset{O-S-O}{||}}\right]^{2-}\leftrightarrow\left[\overset{O}{\underset{O-S=O}{|}}\right]^{2-}$$

Each structure is the same except the location of the double bond is changed.

EXPERIMENT 10   BONDING AND MOLECULAR GEOMETRY

VSEPR Theory
Three dimensional molecular shapes of compounds are predicted using the Lewis structure and VSEPR (Valence Shell Electron Pair Repulsion) theory. The hypothesis of this theory is that the most stable structure of a molecule is one in which the repulsions between the pairs of electrons on the central atom is minimized (i.e., the electrons are as far apart from each other as possible). Both bonding and nonbonding (lone pair) electrons of the central atom are included. Table 10.1 gives the arrangement of electron pairs and molecular geometry for different combinations of electron pairs. <u>It is important to note that for arrangement of pairs purposes only, double and triple bonds are treated as "one" bonding pair in VSEPR Theory.</u>

Also note in the table the placement of lone pairs in the trigonal bipyramidal and octahedral arrangements. The lone pairs are placed on the atom where they are as far apart from other lone pairs or atoms as possible. For example, in the T-shaped figure of AX_3E_2, both lone pairs are placed in the equatorial position ($120°$ from each other) and not one in the axial and one in the equatorial position ($90°$ from each other). Lone pairs also affect the bond angles between atoms in a molecule. Lone pairs take up a great deal of space and subsequently push atoms in the same plane closer together. This explains the measured $104.5°$ angle between O and H in water molecules, different from the $109.5°$ angle in a tetrahedral molecule without lone pairs.

Table 10.1: Arrangement of Electron Pairs and Molecular Geometry

Electron Pairs			Arrangement of Pairs	Molecular Geometry		Bond Angles with Central Atom	Example
Total Pairs	Bonding Pairs	Lone Pairs					
2	2	0	Linear	Linear	$X—A—X$	180°	BeF_2
3	3	0	Trigonal Planar	Trigonal Planar		120°	BF_3
	2	1		Bent		< 120°	SO_2
4	4	0	Tetrahedral	Tetrahedral		109.5°	CH_4
	3	1		Trigonal Pyramidal		< 109.5°	NH_3
	2	2		Bent		~ 104.5°	H_2O

Electron Pairs			Arrangement of Pairs	Molecular Geometry	Bond Angles with Central Atom	Example
Total Pairs	Bonding Pairs	Lone Pairs				
5	5	0	Trigonal Bipyramidal	Trigonal Bipyramidal	90°, 120°	PCl_5
	4	1		Seesaw	90°, < 120°	SF_4
	3	2		T-Shaped	90°	ClF_3
	2	3		Linear	180°	XeF_2
6	6	0	Octahedral	Octahedral	90°	SF_6
	5	1		Square Pyramidal	90°	IF_5
	4	2		Square Planar AX_4E_2	90°	XeF_4

Polarity

Many physical properties, such as melting point, boiling point, and solubility are affected by the polarity of a molecule. <u>Bond</u> polarity is determined by calculating the dipole moment of the bond. The dipole moment (μ) is a quantitative measure equal to the product of the charge, Q, and the distance, r, between the charges in a bond. A general correlation of the dipole moment of a bond is based upon the electronegativities of the elements in the bond. As noted previously, electronegativity is the ability of an atom to attract electrons toward itself in a bond so that the electrons spend more time in the area of the more electronegative element. Electronegativity increases from left to right across a period in the periodic table. In HF, the electronegativity of H is 2.20 and that of F is 3.98. It is represented as

$$\longmapsto$$
$$\text{H} \quad \text{F}$$

where the crossed arrow shows a charge separation towards the more electronegative element. The electron from the H will be drawn to the F atom (i.e., both electrons will spend more time in the vicinity of the F atom), and the molecule HF has a dipole moment towards the F atom.

There is a distinction between <u>bond</u> polarity (the polarity of an individual bond between two elements) and <u>molecular</u> polarity (the overall polarity of the entire molecule which includes all bonds). If a dipole moment exists for a bond, the <u>bond</u> is polar. To determine the polarity of a molecule, the dipole moments of the individual bonds must be considered collectively. Knowing the molecular geometry of the molecule, the bond dipoles can be added as vector quantities to determine the overall dipole moment of a compound. In symmetrical molecules where terminal elements are the same such as CO_2, BCl_3, and CCl_4, the bond dipoles cancel to give a nonpolar molecule. In asymmetrical molecules (diatomic molecule of two different atoms or polyatomic molecules of three or more different atoms), the bond dipoles usually do not cancel and the molecule is polar. Additionally, molecules containing lone pairs on the central atom are polar, unless they are the linear form of the trigonal bipyramidal shape or square planar geometries, which are nonpolar. Double and triple bonds are treated as single bonds when determining molecular polarity. Some polar molecules are HCN, CH_3Br, and H_2O.

Valence Bond Theory and Hybridization of Atomic Orbitals

Although VSEPR Theory is accurate in predicting molecular geometries, it does not explain how the bonds are formed. Valence bond theory assumes that electrons in a molecule occupy atomic orbitals (s, p, d) of individual atoms. Covalent bonds result from the overlap of these atomic orbitals, where the electrons share a common region in space. In order to bond, the atomic orbitals need to be hybridized to form hybrid orbitals. This is necessary because the atomic orbitals of an atom have two or more nonequivalent (i.e. s, p, or d) orbitals. They must combine to form equivalent type orbitals in preparation for covalent bond formation. This mixing of atomic orbitals creates new identical hybrid orbitals.

As a rule, *n* atomic orbitals form *n* hybridized orbitals.

Atomic orbitals of the central atom	Number of hybridized orbitals	Hybridization of the central atom	Arrangement of Pairs	Example
s, p	2	sp	Linear	BeF_2
s, p, p	3	sp^2	Trigonal planar	BF_3
s, p, p, p	4	sp^3	Tetrahedral	CH_4
s, p, p, p, d	5	sp^3d	Trigonal bipyramidal	PCl_5
s, p, p, p, d, d	6	sp^3d^2	Octahedral	SF_6

For example, consider BF_3. The boron has three electrons in its valence shell, and therefore three atomic orbitals. Two electrons will occupy the s orbital and one electron will occupy one of the p orbitals. However, this shows that only one electron in the p orbital is available for bonding and the bond angles between B and F would be 90°. It is known that BF_3 is trigonal planar with bond angles of 120°. Therefore, the s and p orbitals of boron are hybridized (made identical) as three sp^2 orbitals. The s electrons become unpaired and each of the newly unpaired electrons is promoted to one of the three sp^2 hybridized orbitals. This now predicts three identical bonds which are at 120° angles. The three F atoms can now bond with the boron atom in three identical bonds.

Boron:
Ground state:

$$\uparrow\downarrow \quad \uparrow \quad - \quad -$$

2s   2p 2p 2p

Hybridized state:

$$\uparrow \quad \uparrow \quad \uparrow \qquad -$$

sp^2 sp^2 sp^2 2p

The three unpaired electrons are now ready for bonding with three fluorine atoms.

As another example, predict the hybridization of oxygen in water. H_2O is a tetrahedral molecule with a bent molecular geometry. The oxygen atom has six valence electrons, two occupying the s shell (paired) and four occupying the p shell.

Oxygen:
Ground state:

$$\uparrow\downarrow \quad \uparrow\downarrow \quad \uparrow \quad \uparrow$$

2s   2p 2p 2p

The four atomic orbitals of oxygen (s, p, p, p) are then hybridized to make four identical hybridized sp^3 orbitals.

Hybridized state: ↑ ↑ ↑↓ ↑↓   The two unpaired electrons are now ready for bonding with two hydrogen atoms.

sp^3 sp^3 sp^3 sp^3

Two orbitals have one electron in them, each ready to bond with a hydrogen atom, and the other two orbitals contain paired electrons, which represent the lone pairs on the oxygen. Therefore, oxygen is sp^3 hybridized.

N.B. Write the pre-lab up with the following three sections:
- **Purpose**
- **Method**
- **Summary (in table form) of Table 10.1.**

<u>Procedure</u>
Lab Sheet A

Using the molecular model kits, construct the molecules outlined on the index cards and complete the table. Leave the molecule on the card on top of the table for the instructor to check.

- Put ions in brackets with the charge as a superscript
- For ions, put the Lewis structure in brackets with the charge as a superscript on the right outside of the bracket. Shade in the molecular polarity box.
- Build each model and complete its line in the data sheet at the same time

Use the following information to construct the molecules:

1. <u>Atoms</u>: use the appropriate-holed atom for the arrangement of pairs:

Octahedral:	6-holed, yellow
Trigonal bipyramidal:	5-holed, purple
Tetrahedral:	4-holed (black, blue, and red)
Trigonal:	3-holed (blue)
Linear:	2-holed, red

One holed atoms are included to represent terminal atoms: white, green, light green, and orange.

2. <u>Lone pairs</u>: use the small white half sphere and no bond link.

3. <u>Bonds</u>:
   a. Represent single bonds with the short rigid grey link.
   b. Represent one double bond using two longer flexible gray links.
        Use a 4-hole atom, regardless of the arrangement of pairs.
   c. Represent one triple bond using three longer flexible gray links.
        Use a 4-hole atom, regardless of the arrangement of pairs.

4. Use different colored balls to represent different atoms in a molecule.

5. For organic molecules, if there appears to be more than one central element, pick one of the atoms to be a central atom.

Lab Sheet B
Complete the table according to the given cards.

Lab Sheet C
Complete the page.

Lab Sheet A: Drawing Lewis Structures, Making Models and Predicting Molecular Geometry Properties

Box # _______________

Molecule	Lewis structure	VSEPR Bonding Pairs @ Central Atom	Lone Pairs @ Central Atom	Arrangement of Pairs	Molecular Geometry Name and 3D Structure	Bond Angles between the Atoms	Bond Polarity	Molecular Polarity (**neutral molecules only**)	Hybridization of Central Atom

Molecule	Lewis structure	VSEPR Bonding Pairs @ Central Atom	Lone Pairs @ Central Atom	Arrangement of Pairs	Molecular Geometry Name and 3D Structure	Bond Angles between the Atoms	Bond Polarity	Molecular Polarity (**neutral molecules only**)	Hybridization of Central Atom

Molecule	Lewis structure	VSEPR Bonding Pairs @ Central Atom	Lone Pairs @ Central Atom	Arrangement of Pairs	Molecular Geometry Name and 3D Structure	Bond Angles between the Atoms	Bond Polarity	Molecular Polarity (**neutral molecules only**)	Hybridization of Central Atom
Complete for one central atom									

Lab Sheet B

Copy the two structures from each card onto this paper.   Draw in lone pairs for each molecule. Calculate the formal charges for each element and select the best structure. State why the selected structure is preferred.

1. Card Number:______________________

Why preferred:

2. Card Number:______________________

Why preferred:

Lab Sheet C:

Draw the following molecules showing resonance between the structures:
> Use correct resonance structure notation (brackets and arrows)
> If the structure is an ion, include the charge as a superscript outside the bracket

1. HCO_2^- (two structures)

Arrangement of Pairs	Structure Geometry	Bond Angles Between Atoms

2. ClO_3^- (three structures)   *To minimize formal charge, Cl has an expanded octet. It has two sets of double bonds and a lone pair.*

Arrangement of Pairs	Structure Geometry	Bond Angles Between Atoms

3. N_2O, arranged in the order NNO (three structures)

Arrangement of Pairs	Structure Geometry	Bond Angles Between Atoms

EXPERIMENT 10   BONDING AND MOLECULAR GEOMETRY

<u>Conclusion/Error Analysis</u>

Complete the following sentences.

1. Lone pairs go in the _________________ position of trigonal bipyramidal molecules.

2. Lone pairs go in the _________________ position of octahedral molecules.

3. Lone pairs usually make a molecule polar. However the two exceptions are the following:

    _______________ molecular geometry of the _______________ arrangement *and*

    _______________ molecular geometry of the _______________ arrangement

4. In determining formal charge, the preferred structure is the one in which the negative charge is placed on the most _________________ element.

5. For a compound that can be drawn two different ways using the same skeletal structure, but bond types are different and one contains an expanded octet, the correct structure can either be selected based on the fact that either all formal charges are ___________ OR the compound obeys the _________________.

6. The hybridization of the central atom of a tetrahedral molecule is _____________, while the hybridization of the central atom of an octahedral molecule is __________.

144

EXPERIMENT 10   BONDING AND MOLECULAR GEOMETRY

HOMEWORK EXERCISES   Name _______________________

1. Briefly define the following terms:
   <u>Lewis dot structure</u>

   <u>Formal charge</u>

   <u>Resonance</u>

   <u>Electronegativity</u>

   <u>Polar bonds</u>

   <u>Polar molecule</u>

   <u>Non polar molecule</u>

   <u>Lone Pairs</u>

   <u>Atomic orbitals</u>

   <u>Hybridization</u>

2. What are the five possible arrangements of atoms in a molecule?

3. How does a lone pair affect the bond angles within a molecule?

4. Draw the Lewis structures for the following molecules: (<u>include all electrons</u>)

CBr_4		NH_3	
BCl_3		CO_2	
CCl_3Br		H_2O	

5.   Draw the structure and calculate the formal charge on each atom of NO_2^-.

O   N   O

6.   Complete the following table:

	Arrangement of Pairs	Molecular Geometry	Hybridization
3 bonding pairs and 1 lone pair			
3 bonding pairs and 0 lone pairs			
3 bonding pairs and 2 lone pairs			
2 bonding pairs and 0 lone pairs			
4 bonding pairs and 2 lone pairs			

7.   List the polarity of the bonds and the molecules (refer to Question 4 for the Lewis structures):

Molecule	Bond	Bond Polarity (Polar or Nonpolar)	Molecule Polarity (Polar or Nonpolar)
CBr_4	C–Br		
CCl_3Br	C–Cl		
	C–Br		

8.   Why must atomic orbitals be hybridized before bonding?

Technique Lab – Chem 1114

This lab gives experience in using and properly reading common Chem 1114 equipment and constructing a graph.

Name_________________________________   Partner_________________________________

Date_________________________________   Section_________________________________

<u>Part 1: Reading the barometer and converting units.</u>

Read the red scale (mm Hg) of the barometer in the lab. Since each subdivision of this scale is 1 mm Hg, estimate to the tenths place.

Record the value: ____________________ mm of Hg

Number of significant figures in this value: ____________

Many units are used for pressure. Using the following conversion factors and dimensional analysis, convert the lab barometer reading to the necessary units.

1 inch Hg = 25.4 mm Hg

1 torr = 1 mm Hg

1 atm = 760 mm Hg

1 atm = 101,325 Pa

<table>
<tr><td>

<u>Important Point! Read Carefully!</u>
These values are exact numbers. Exact numbers are numbers obtained from definitions (conversion factors) or by counting the number of objects in a group (such as "12" are in a dozen). *Exact numbers have an infinite number of significant figures.* Therefore, the number of significant figures in a converted answer below is dependent upon the number of significant figures in the barometer reading.

</td></tr>
</table>

<u>SHOW ALL WORK USING DIMENSIONAL ANALYSIS!</u>

Barometric pressure, torr: ____________________

Barometric pressure, inches of Hg: ____________________

Barometric pressure, atm: ____________________

Barometric pressure, Pa: ____________________

Part 2: Use of the thermometers. Calculation of °C and K and % error.

Important Points:
1. USE UNITS AT ALL TIMES!!!
2. The numbers in parenthesis are the number of significant figures.

1. Using a 0 – 100 °C thermometer, determiner the temperature of the air in the lab. *This temperature should be read to the nearest 0.1 degree.*	Temperature: (tenths)
2. Determine the temperature of the air in the lab using a 0 – 50°C thermometer. *This temperature should be read to the nearest 0.01 degree.*	Temperature: (hundredths)
3. When cooling a solution at 10°C to 1°C, which thermometer should be used if very precise temperature changes need to be recorded?	
4. Which thermometer should NEVER be used to determine the temperature of water that will be heated to boiling? Why?	
5. Convert the 0 – 50°C thermometer reading from #2 to °F. $$\left(\frac{9}{5}\right)°C + 32 = °F$$	Temperature: (4)
6. Convert the 0 – 50°C thermometer reading from #2 to K. $$K = °C + 273.15$$	Temperature: (5)
7. Assuming that the theoretical value of frozen water is 32.00 °F, and an experiment recorded that it was 31.85°F, calculate a % error of the temperature. $$\% \text{ error} = \frac{\text{experimental value - theoretical value}}{\text{theoretical value}} \times 100$$	% Error: (2)
8. Based on the % error, is the thermometer used in this experiment reliable? (Should be ± 5%)	

Note: A positive % error means the experimental value is larger than the theoretical value. A negative % error means the experimental value is smaller than the theoretical value. It gives a magnitude of the error, but does not indicate what type (systematic or random) of error.

Name_________________________   Partner_________________________

Date_________________________   Section_________________________

<u>Part 3: Use of the buret, pipet, and pH meter. Construct a proper graph. Use units!!!</u>

<u>Titration of Acetic Acid with NaOH</u>

1. Calibrate the digital pH meter according to the instructions in the Appendix.

2. Obtain approximately 50 mL of standardized NaOH in a clean and dry 100 mL beaker. Normalize the 50 mL buret with this solution then fill it.

3. With an autodispenser, dispense exactly 50.0 mL of an acetic acid solution to a 250 mL beaker using the following method: First, place a waste beaker under the delivery spout and gently pull the autodispenser barrel upwards. Touch any drops at the end of the delivery spout to the side of the beaker. Next, place the 250 mL beaker under the spout and gently push the autodispenser barrel down. Add 3-4 drops of phenolphthalein. Swirl to mix.

4. Unscrew the pH probe from the storage container (if applicable). Leave this top on during the titration. Rinse the pH electrode with DI water, gently shake it off and blot with a paper towel. Place the pH electrode in the beaker containing the acid solution. The electrode can remain in the beaker for the remainder of the experiment, and can be used as a stirring rod.

5. Record the initial pH and initial buret reading in the laboratory report sheet. Subtract the <u>initial</u> buret reading from each subsequent buret reading to obtain a running total of the mL of NaOH added.

6. Add the first increment of about 1.00 mL of NaOH solution to the beaker containing the acetic acid. Gently mix the solution with the electrode and wait for the "S" to appear. Record the pH and buret volume in the laboratory report sheet.

7. Continue to add volumes of NaOH, about 1.00 mL at a time, reading and recording the pH and buret readings, until the titration curve shows little pH change (less than 0.05 units) with added NaOH and the pH is above 12 for two readings.

8. Rinse the electrode with DI water and replace it in its container.

9. Drain the remaining NaOH to the sink, rinse the buret with DI H_2O, flip the buret and open the stopcock. All waste should go to the drain.

10. Using the data obtained, construct a proper graph. Plot the data in a landscape orientation. Label all axes correctly and include a descriptive title. Connect the data points in a smooth curve. Refer to the Appendix for guidelines.

<u>Always keep the pH electrode immersed and upright when not in use.</u>
<u>Notify the instructor if more 3 M KCl solution needs to be added to the storage container.</u>

LABORATORY REPORT SHEET –PART 3

Name _________________________________ Partner _________________________________

Titration Data of Acetic Acid ($HC_2H_3O_2$) and NaOH

Buret Reading, mL	Total mL NaOH added	pH		Buret Reading, mL	Total mL NaOH added	pH
*	0.00 mL					
				When the titration is complete, drain the buret. Fill with DI water and drain to the sink or waste beaker. Invert the buret, clamp to the buret holder and open the stopcock.		

*   This initial reading should be subtracted from subsequent buret readings (column 1) to obtain the total mL of NaOH added (column 2).

Graph the data above (columns 2 and 3). Refer to the Appendix for proper graphing techniques.

EXPERIMENT 11
MOLAR MASS DETERMINATION BY FREEZING POINT DEPRESSION

The freezing point depression caused by the dissolution of a solute in a solvent will be used to determine the molar mass of the solute.

Key Chemical Reactions: None

Key Mathematical Equations: $\Delta T_f = K_f m$

Discussion

Freezing point depression, boiling point elevation, vapor pressure lowering, and osmotic pressure are all colligative properties – properties that are solely dependent upon the number of solute particles (ions, atoms, or molecules) present in a solution, and not the identity of the solute.

The addition of a non-volatile solute to a solvent makes the resulting solution have a lower vapor pressure than the solvent at all temperatures. The decrease in the vapor pressure causes a depression of the freezing point below that of the pure solvent and the elevation of the boiling point. The phase diagram in Figure 11.1 illustrates the freezing point depression and boiling point elevation of a substance when a solute is added.

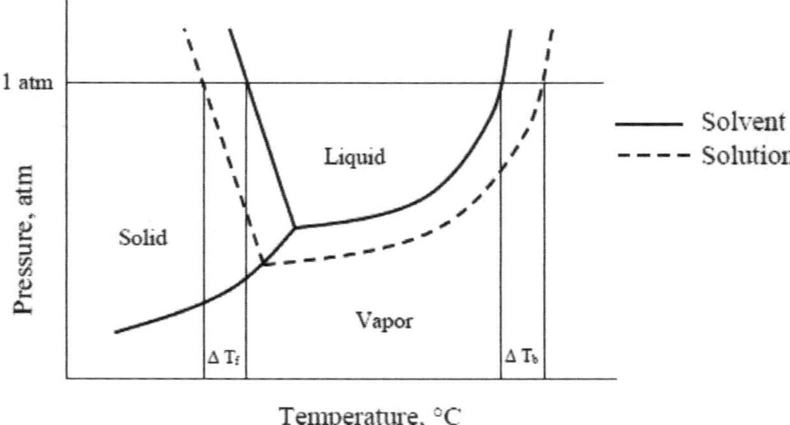

Figure 11.1
Phase Diagram of a Solvent and Solution

Freezing is a phase transition from a disordered state (liquid) to a more ordered state (solid). During this transition, energy is removed from the system. When comparing the freezing point of a solution and a pure solvent, more energy must be removed from a solution because a solution has greater disorder than a pure solvent. Therefore, the solution will have a lower freezing point than the pure solvent.

For dilute solutions, the decrease in the freezing point of a solution from that of the pure solvent, ΔT_f, is directly proportional to the molality of the solution, expressed in the following relationship:

$$\Delta T_f = K_f m$$

Where

K_f (units: °C/m), is the molal freezing point depression constant, a function of the solvent only, and

m is the molality of the solution, moles of solute per kg of solvent.

A similar equation expresses the boiling point elevation:

$$\Delta T_b = K_b m$$

where ΔT_b is the boiling point elevation and K_b is the molal boiling point constant. Values of the constants K_f and K_b for several solvents are given in Table 11.1.

Table 11.1
Molal Freezing and Boiling Point Constants for Various Solvents

Substance	$K_f\ (^{\circ}C/m)$	$K_b\ (^{\circ}C/m)$
Benzene	5.12	2.53
Carbon tetrachloride	29.8	5.03
Cyclohexane	20.0	2.79
Glacial acetic acid	3.90	2.93
Naphthalene	6.85	5.80
Tertiary butyl alcohol	8.35	-
Water	1.853	0.512

In this experiment, the freezing point of a pure solvent is measured. A known mass of solute is added to a given mass of the solvent and the freezing point of the resulting solution measured. The molar mass of the solute is then calculated from these data.

Example:
Calculate the molar mass of a solute if a freezing point depression of 1.34 °C is measured when 0.2409 g of the solute is added to 10.0000 g of tertiary butyl alcohol.

First, calculate the molality of the solution.

$$\Delta T_f = K_f m$$

$$1.34\ ^{\circ}C = (8.35\ \frac{^{\circ}C}{m})\, m$$

$$m = 0.161\ \frac{\text{moles solute}}{\text{kg solvent}}$$

Using dimensional analysis, multiply this value of molality by the mass of the solvent to obtain the moles of the solute.

$$0.161 \ \frac{\text{moles solute}}{\text{kg solvent}} \ \text{x} \ 0.0100000 \ \text{kg solvent} = 1.61 \ \text{x} \ 10^{-3} \ \text{moles solute}$$

The molar mass can then be calculated since the mass of the solute was measured.

$$\frac{0.2409 \ \text{g solute}}{1.61 \ \text{x} \ 10^{-3} \ \text{moles solute}} = 150. \ \text{g/mole}$$

An additional amount of solute may be added to the solution, and the freezing point will be even lower. In these cases, the total mass of the solute must be accounted for, and the difference in temperature is the freezing point depression between the solvent and final solution.

In any freezing point depression experiment, the pure solvent's freezing point should be measured experimentally, and not taken from literature values. The solvent may not be perfectly pure, in which case its freezing point will usually be slightly low. Also, the thermometer may be inaccurate by as much as a degree or two. However, it is not the accuracy of the freezing point of solvent or solution which is important but the <u>difference</u> between these two readings. Therefore, many errors will be minimized if the same thermometer and the same solvent are used throughout the experiment.

The method used in this experiment involves cooling the solvent or solution in an ice bath and recording the temperature and time data as the liquid cools. As the solid is formed, heat is evolved (heat of fusion), and this can offset the cooling process. The accurate measurement of the freezing point is a key part of the procedure in this experiment. There are three possible types of cooling curves that may be encountered: a pure substance, a solution, and a pure substance or a solution that supercools. Each of these has a specific method to determine the freezing point.

Pure substance freezing point determination:
For a completely pure substance under constant pressure, the solid and liquid phases coexist in equilibrium, and the temperature at the freezing point should remain constant. The substance will cool to the freezing point, and the temperature will be constant (slope of zero) until all of the substance is frozen. The temperature that corresponds to this zero slope line is the freezing point of the substance.

Solution freezing point determination:
In a solution, the measured freezing point will usually decrease slowly as crystallization takes place. The reason for this is that only the solvent crystallizes out; the solute remains in the liquid phase. This causes the remaining liquid solution to become more concentrated, thereby decreasing the freezing point more and more. The freezing point of a solution is indicated by a change in the slope (the first point of inflection) on the cooling curve.

Supercooling freezing point determination (bottom curve of Figure 11.2):
In many cases, however, as the temperature drops, the liquid often "supercools", where it exists in a metastable liquid state below the freezing point before it crystallizes. (This is shown as a big "dip" in the cooling curves of Figure 11.2.) When supercooling occurs in a *pure liquid*, the temperature of the liquid drops below the freezing point and then increases rapidly to reach a constant value that persists until all of the liquid is frozen to a solid. When supercooling occurs in a *solution*, the temperature of the solution also drops below the freezing point and then increases rapidly. During freezing, the temperature will continue to drop because the concentration of the solution increases as the pure solvent freezes out. Figure 11.2 depicts the temperature vs. time plot when supercooling takes place in both these situations. The best estimate of the freezing point is obtained by extrapolating the linear portion of the temperature vs. time curve just after the supercooling dip back to the point of intersection with the original cooling curve. This is shown in Figure 11.2.

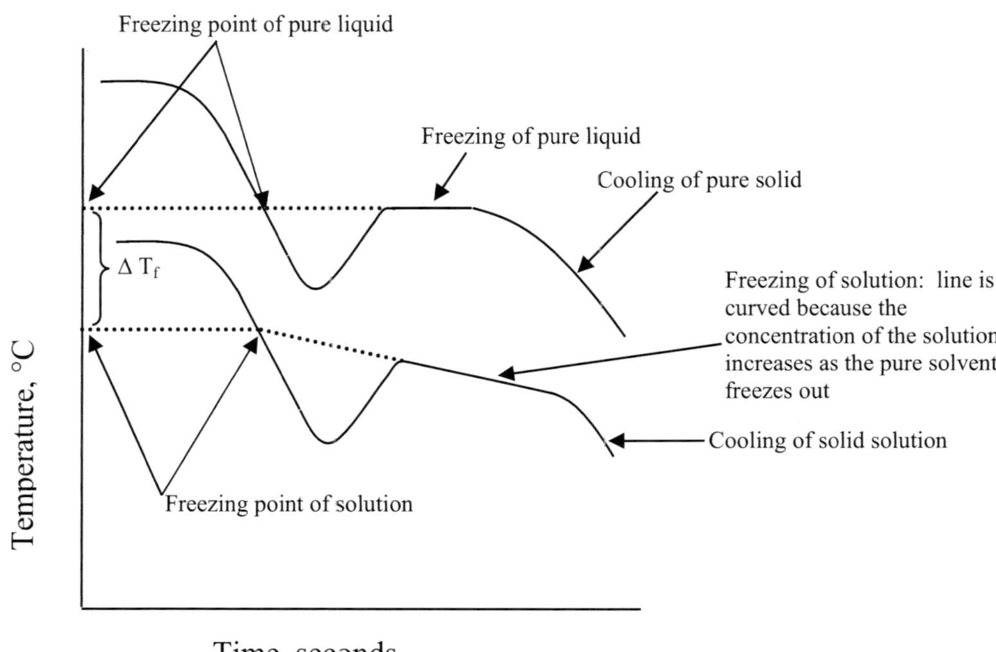

Figure 11.2
Temperature vs. Time Cooling Curve for a Pure Solvent and a Solution

Chemicals
    Glacial Acetic Acid CH_3COOH
    Various Organic Solids

Waste
    All waste should be disposed of in the laboratory hazardous waste container.

<u>Procedure</u>
1. Dry both test tubes completely.

2. Begin to heat tap water in a 400 mL beaker on a hot plate. Set on the lowest setting.
    Add water as necessary throughout the experiment to maintain the water level at least
    half full.

3. Assemble the apparatus as follows: Fill a 600 mL beaker with tap water to the 400
    mL mark, and place it on the base of a ring stand. Attach a test tube clamp and a
    thermometer clamp above it to the ring stand.

4. Weigh a clean and dry test tube together with the foam support.

5. Accurately weigh approximately 0.5 g of the unknown, and then carefully transfer it
    from the weighing paper to weighed test tube. Tap the test tube gently to remove any
    solid from the test tube walls.

6. Rezero the balance and weigh the test tube, foam support, and unknown. Set this test
    tube, unknown, and foam support aside for use at a later step.

7. Pour 1 – 1½ inches of glacial acetic acid into another clean and dry test tube. An
    exact measurement is not necessary.

8. Clamp the test tube of acetic acid to the test tube clamp on the ring stand and lower
    the test tube into the water in the 600 mL beaker so that the level of the acetic acid is
    well below the level of the water.

9. Insert a metal stirrer (loop side down) into the test tube. Insert a thermometer with
    tenth– degree divisions in the thermometer clamp and then lower the thermometer
    into the test tube. Insure that the bottom of the thermometer is not touching the
    bottom of the test tube and is immersed in the acetic acid. Clamp the thermometer
    into the thermometer clamp.   Insert a thermometer with one–degree divisions into
    the water in the 600 mL beaker to monitor the temperature of the bath.

10. Begin the cooling by adding ice to the 600 mL beaker to bring the temperature of the
    bath down to about 14°C. Maintain it at this temperature by administering warm
    water or ice. Do not cool it below 10°C or the freezing point will be inaccurate. DO
    NOT add ice with the same hand that is stirring the solution.

11. Begin collecting time and temperature (from the thermometer in the test tube) data
    when the acetic acid is 17°C. One partner should read the temperature of the solvent
    and stir continuously in a gentle up and down motion. The other partner should read
    the time, record data, and visually inspect the solution. The solvent temperature will
    drop rapidly. As freezing occurs, a large amount of heat will be generated. The
    temperature of the solvent will rise sharply in a matter of seconds. Record the
    maximum temperature during this time. Continue to stir and collect data for one
    minute after the readings stabilize (less than 0.1 °C change).

Note: the solution is "frozen" when it appears slushy and translucent or opaque. It will not be completely solid; the stirrer should still be able to move. The solution must freeze with this appearance to obtain the best results. If the solution has not frozen, continue to add more ice to the ice bath.

12. Unclamp the thermometer and the test tube and then remove the test tube of acetic acid <u>with the stirrer and thermometer still in the test tube</u>. Place the test tube, stirrer, and thermometer assembly in the warm water and heat it until the acetic acid is completely liquid. Some stirring may be necessary.

13. When no solid acetic acid remains, remove the test tube from the water and wipe the outer walls off with a paper towel. Remove the stirrer and thermometer and carefully transfer all of the acetic acid to the test tube containing the unknown. Make sure no solid is on the walls of the test tube.

14. Weigh the test tube with the unknown and acetic acid together with the foam support.

15. If the solid is not completely dissolved in the acetic acid, gently warm it in the warm water to dissolve.

16. Dispose of the water in the 600 mL beaker and replace it with 400 mL of tap water.

17. Clamp the test tube to the test tube clamp on the ring stand and lower the test tube into the water in the 600 mL beaker so that the level of the solution is well below the level of the water.

18. Wipe off and dry the metal stirrer and insert it into the test tube. Wipe off and dry the thermometer and lower it into the test tube, making sure the bulb of the thermometer is not touching the bottom of the test tube and it is immersed in the solution. Clamp the thermometer to the thermometer clamp. The thermometer with 1° divisions should remain in the 600 mL beaker to monitor the temperature of the bath.

19. Begin the cooling by adding ice to the 600 mL beaker to bring the temperature of the bath between 10 – 12°C. Maintain it at this temperature by administering warm water or ice. Do not cool it below 8°C or the freezing point will be inaccurate.

20. Collect temperature and time data as before, reading the thermometer in the test tube when the test tube contents are 17°C. One partner should read the temperature of the solvent and stir continuously. The other partner should read the time, record data, and visually inspect the solution. The solvent temperature will drop rapidly. As freezing occurs, a large amount of heat will be generated. The temperature of the solvent will rise sharply in a matter of seconds. Record the maximum temperature during this time. Continue to stir and collect data for one minute after the temperature stabilizes.

21. Unclamp the thermometer and the test tube. Remove the test tube with the stirrer and the thermometer assembly from the 600 mL beaker. Place the apparatus in the beaker containing warm water to melt the acetic acid solution. Some stirring may be necessary.

22. Weigh an additional 0.5 g of the unknown and record the mass.

23. When no solid acetic acid remains, remove the test tube from the water and wipe the outer walls off with a paper towel. Lift the thermometer out of the solution, touch it to the side of the test tube so that any solution on the thermometer drains down into the test tube, and remove the thermometer. Lift the stirrer out of the solution, touch it to the side of the test tube, and remove the stirrer. Carefully transfer all of the unknown to the solution. It may be helpful to tilt the test tube so that all of the unknown reaches the solution, and none remains on the walls of the test tube.

24. Place the test tube in the warm water to completely dissolve the solid unknown. Some stirring may be necessary.

25. Dispose of the water in the 600 mL beaker and replace it with 400 mL of tap water.

26. Clamp the test tube to the test tube clamp on the ring stand and lower the test tube into the water in the 600 mL beaker so that the level of the acetic acid is well below the level of the water. Insert the thermometer and clamp it as done previously. Insert the stirrer.

27. Begin the cooling by adding ice to the 600 mL beaker to bring the temperature of the bath between 10 – 12°C. Maintain it at this temperature by administering warm water or ice. Do not cool it below 8°C or the freezing point will be inaccurate.

28. Collect temperature and time data as before. Start recording data when the contents of the test tube are 17°C. Again, as freezing occurs, a large amount of heat will be generated. The temperature of the solvent will rise sharply in a matter of seconds. Record the maximum temperature during this time. Continue to stir and collect data until for one minute.

29. Unclamp the thermometer and test tube and remelt the contents of the test tube in the warm water. Dispose of the acetic acid solution in the laboratory waste container. Clean all glassware well.

     *** It may be necessary to rinse the test tubes out with small amounts of acetone.
     The acetone should also be disposed of in the waste container. ***

30. Plot the three cooling curves (all on one graph) to graphically determine the freezing point of each run. Scale the temperature axis appropriately (~12° – 17°C) so that the graph is as large as possible.

LABORATORY REPORT SHEET

Name____Date________________

Partner ________________________________

Time (seconds)	Run 1 Pure glacial acetic acid Temperature, °C	Run 2 First addition of unknown Temperature, °C	Run 3 Second addition of unknown Temperature, °C
0			
15			
30			
45			
60			
75			
90			
105			
120			
135			
150			
165			
180			
195			
210			
225			
240			
255			
270			
285			
300			
315			
330			
345			
360			
375			
390			
405			
420			
435			
450			
465			
480			

Unknown Number ________________________

Include units and correct significant figures on all data and calculations

<u>Mass Data</u>

Mass of empty test tube A and foam support ________________

Mass of test tube A, foam support, and unknown ________________

  Mass of unknown, first addition ________________

Mass of test tube A, foam support,
 unknown and acetic acid ________________

Mass of acetic acid ________________

  Mass of unknown, second addition ________________

<u>Temperature Data</u>

Freezing point of pure acetic acid ________________

Freezing point of acetic acid and unknown ________________ ΔT ________________

Freezing point of acetic acid and both unknowns ________________ ΔT ________________

FP Depression Equation ________________________ K_f ________________________

<u>Calculation of molar mass of solute, Runs 1–2:</u>

<u>Calculation of molar mass of solute, Runs 1–3:</u>

<u>Average molar mass of solute:</u>

Error Analysis

1. If the unknown solute did not completely dissolve, how would this affect the calculated molar mass of the solute? (1) Clearly explain why, discussing the impact on the number of particles in solution and the effect on the final freezing point of the solution and the ΔT observed. (3)

2. If the bath was too cold, some of the heat generated during supercooling would be absorbed by the cold bath. How would this affect the calculated molar mass of the solute? (1) Clearly explain why, discussing the effects on the final freezing point of the solution and the ΔT observed. (3)

3. Why is molality used in freezing point depression experiments and not molarity? (3)

4. Technique is extremely important in this experiment to obtain good results. List two techniques, if carefully followed that will improve accuracy. (2)
   (1)

   (2)

EXPERIMENT 11   MOLAR MASS DETERMINATION BY FREEZING POINT DEPRESSION

Name ______________________________

HOMEWORK EXERCISES

1. In an experiment, the freezing point curve of pure benzene is in Trial I. When 0.9584 g of the solute camphor was dissolved in 16.2051 g of benzene, the data in Trial II was obtained. An additional 0.8246 g of camphor was added to the solution and the data in Trial III was obtained.

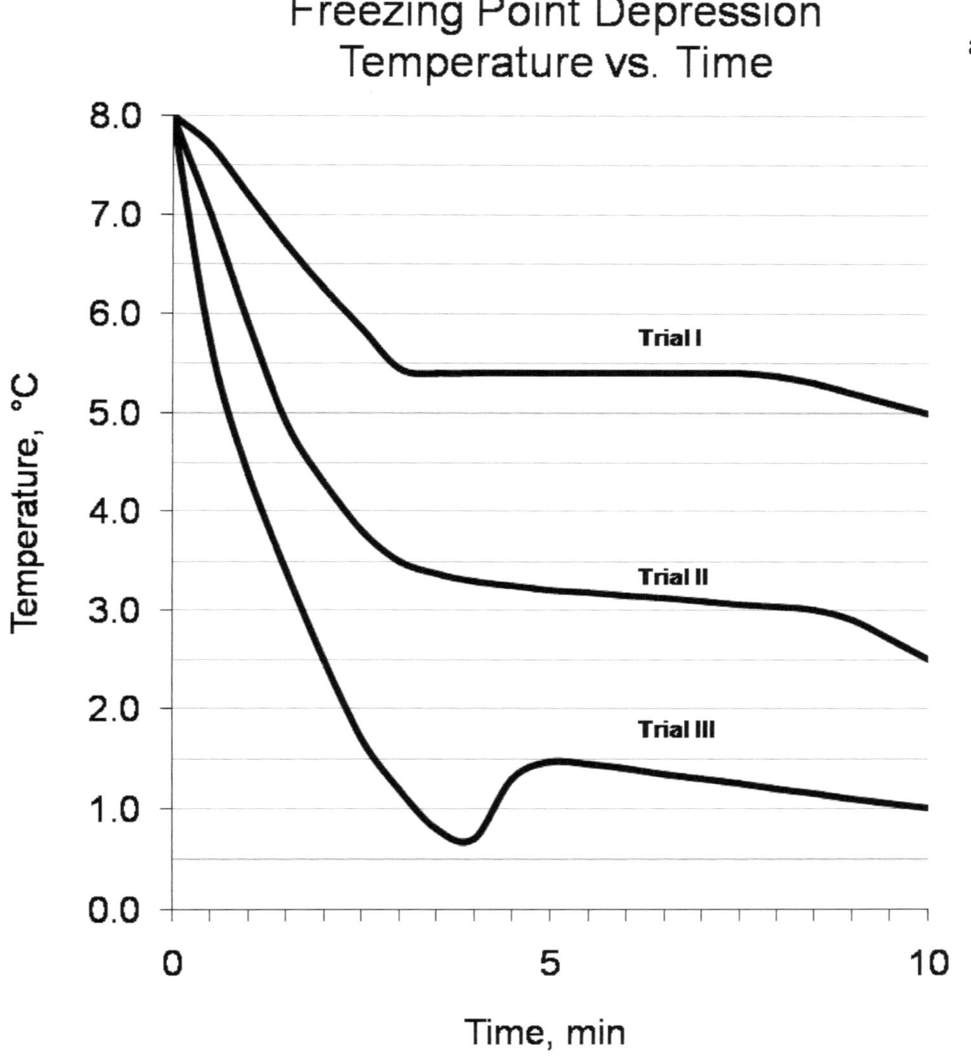

a. Determine the freezing points of runs I, II and III.

   Trial I: __________

   Trial II: __________

   Trial III: __________

b. Calculate the experimental molar mass of camphor using Trials I and II.

Problem 1 is continued on the back

c. Calculate the experimental molar mass of camphor using Trials I and III.

d. Calculate the average molar mass.

2. Why must the freezing point of a solvent be measured in an experiment instead of using the literature value?

3. Why can't molarity be used instead of molality in freezing point depression/boiling point elevation experiments? *Refer to the lecture text.*

4. A 4.1295 g sample of a carbohydrate dissolved in 26.0846 g of pure water gave a solution with a freezing point of $-0.0178°C$. Calculate the molar mass of the carbohydrate. Assume water freezes at $0.0000°C$.

5. In an experiment, the freezing point of 24.6217 g of carbon tetrachloride was determined to be $-21.91°C$. Calculate the freezing point of the solution when 0.4732 g of naphthalene (128.17 g/mole) is added.

EXPERIMENT 12
THE RATE OF CHEMICAL REACTIONS

The initial rate method will be used to determine the rate law of a reaction. The rate will be measured by determining the length of time required to produce a sharp color change in the solution. The rate law for the reaction will be determined, as well as the orders with respect to each of the reactants, the overall order of the reaction, and the rate constant.

Key Chemical Reactions:

$$2\,I^- + S_2O_8^{2-} \rightarrow I_2 + 2\,SO_4^{2-} \qquad \text{(Reaction equation)}$$

$$I_2 + 2\,S_2O_3^{2-} \rightarrow 2\,I^- + S_4O_6^{2-} \qquad \text{(Indicator reaction equation)}$$

Key Mathematical Equations:

$$\frac{\text{rate}_1}{\text{rate}_2} = \left\{ \frac{[S_2O_8^{2-}]_1}{[S_2O_8^{2-}]_2} \right\}^x$$

$$\text{Rate}_{(\text{initial})} = k[A]_0^x[B]_0^y$$

$$C_1V_1 = C_2V_2$$

Discussion

Kinetics is the study of the rates at which chemical reactions take place. The overall rate of a reaction can be measured by observing the rate of disappearance of a reactant or the rate of formation of a product as a function of time. The rate is the speed at which a reaction occurs and is usually expressed in terms of the change in molar concentration per unit of time, e.g., M/s, M/min, etc. For a reaction of the form

$$a\text{A} + b\text{B} \rightarrow c\text{C} + d\text{D}$$

the rate can be expressed by measuring the decrease in the concentration of a reactant or increase in the concentration of a product as a function of time. Mathematically, the rate is expressed as:

$$\text{Rate} = -\frac{1}{a}\frac{\Delta[A]}{\Delta t} = -\frac{1}{b}\frac{\Delta[B]}{\Delta t} = \frac{1}{c}\frac{\Delta[C]}{\Delta t} = \frac{1}{d}\frac{\Delta[D]}{\Delta t}$$

where $\Delta[A]$ is the change in molar concentration of the reactant A in an amount of time Δt. Since the rate is always a positive number, when reactant concentrations are used, a minus sign must be included. The rate of a reaction will decrease as the reaction proceeds since reactants are being consumed. This is shown graphically in Figure 12.1, in a plot of the concentration of a product versus time, where the rate is the instantaneous slope of the curve.

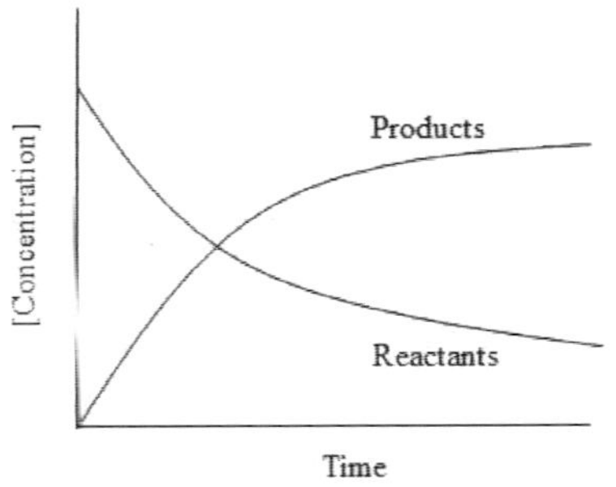

Figure 12.1
Rate of Formation of Products

The plot shows that the concentration of the products increases rapidly early in the reaction, then slowly levels off to a constant value at long times. Similarly, the concentration of reactants versus time shows an initial rapid decrease in concentration at short reaction times, then leveling off at longer times.

The rate of a chemical reaction depends on several factors which include the nature (chemical state, identity, etc.), temperature, and the concentration of the <u>reactants</u>.

The dependence of the rate on concentration can be summarized in a mathematical expression called the rate law. For the reaction aA $+$ bB $\rightarrow$ cC $+$ dD, the rate law is expressed by the following equation:

$$Rate = k[A]^{x}[B]^{y}$$

where
    k is the specific rate constant which depends only on temperature;
    [A] and [B] are the molar concentrations of A and B, respectively; and
    x and y are the orders of reactant A and B, respectively

The orders are constants, and are usually small, positive integers which represent the proportion in which the rate changes with changing concentration. The reactant orders are independent of the stoichiometric coefficients. The magnitudes of x, y, and k must be experimentally determined. The above example reads as follows: The rate of reaction is xth order with respect to A, yth order with respect to B, and $(x + y)$ order overall.

To determine the specific rate law for any reaction, the order with respect to each reactant must first be determined. In the reaction aA $\rightarrow$ products, the rate law is written:

$$Rate = k[A]^{x}$$

One way to determine the order "x" with respect to [A] is to plot a form of the concentration of the reactant ([A], ln[A], or 1/[A]) versus time. Whichever plot gives a straight line determines the order: 0, 1, or 2, respectively.

Figure 12.2 summarizes the reaction orders that give straight lines. Table 12.1 summarizes the rate law, integrated rate law, slope, and half life time for orders 0, 1, and 2.

Zero Order

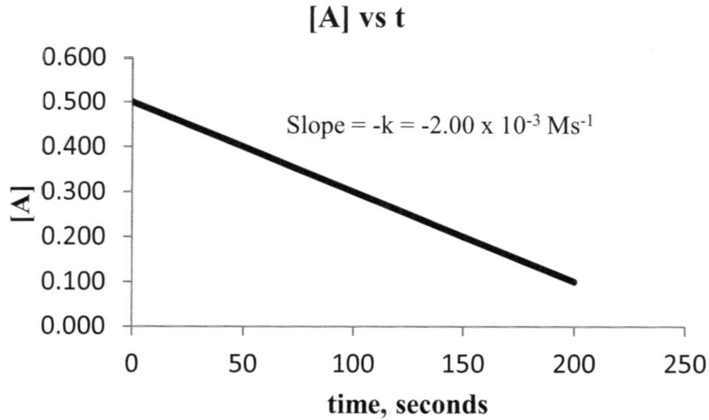

First Order

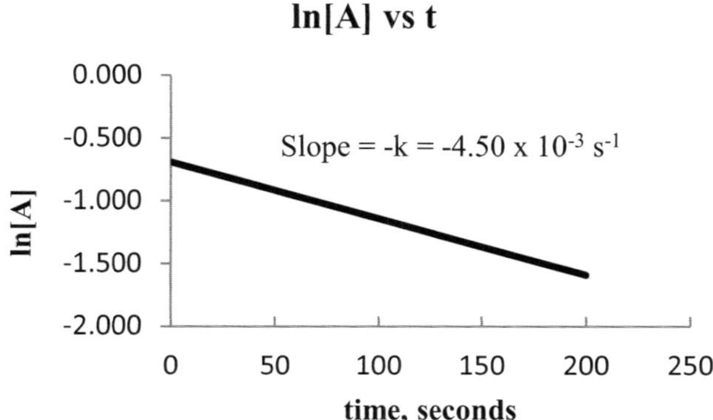

Second Order

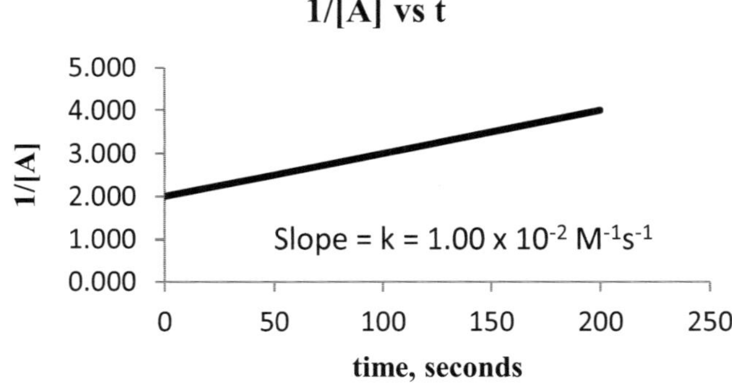

Figure 12.2
Zero, First, and Second Order Plots

Table 12.1. Summary of Order Data

Order	Graph that gives straight line	Rate Law	Integrated Rate Law	Slope	$t_{1/2}$
0	[A] vs t	Rate = k	$[A]_t = -kt + [A]_0$	$-k$	$\dfrac{[A]_0}{2k}$
1	ln [A] vs t	Rate = k[A]	$\ln[A]_t = -kt + \ln[A]_0$	$-k$	$\dfrac{0.693}{k}$
2	1/[A] vs t	Rate = k[A]^2	$1/[A]_t = -kt + 1/[A]_0$	k	$\dfrac{1}{k[A]_0}$

The above method is useful for determining the order of a reactant and the rate law when there is only one reactant, or when monitoring one reactant with other reactants in a large excess. However, when there is more than one reactant and they contribute to the rate law, another method, called the initial rate method, must be used.

Initial Rate Method
The initial rate method is based on the assumption that if the rate of a reaction is measured very soon after mixing the reactants together, such small amounts of reactants will have been consumed that their concentrations are essentially at their initial values and the rate law is given by

$$\text{Rate}_{(initial)} = k[A]_0^x[B]_0^y$$

where $[A]_0$ and $[B]_0$ are the initial concentrations of the reactants A and B. By measuring the initial rates of the reaction at different concentrations of the reactants, the values of k, x, and y can be determined.

For example, for the reaction $A + B \rightarrow$ products, if the concentration of B is held constant and the initial rate doubles when the starting concentration of A is doubled, then the rate is directly proportional to [A], and the constant "x" must have the value of 1. If, in another experiment, [A] is held constant and the initial rate goes up by a factor of four when the starting concentration of B is doubled, then the rate must be proportional to the square of the concentration of B. The value of "y" must then be equal to 2. The rate law then would be:

$$\text{Rate} = k[A][B]^2$$

This method works well when there is a large amount of data and many reactants. For example, in the reaction $aA + bB + cC \rightarrow$ products, the following data was obtained.

Run	[A]	[B]	[C]	Rate, M/s
1	0.100 M	0.200 M	0.200 M	1.20×10^{-2}
2	0.200 M	0.200 M	0.200 M	1.20×10^{-2}
3	0.100 M	0.400 M	0.200 M	4.80×10^{-2}
4	0.100 M	0.200 M	0.400 M	2.40×10^{-2}

To determine the rate with respect to a specific reactant, first select the runs in which that reactant's concentrations change, but the <u>other reactants' concentrations stay the same</u>. For the reactant A, Runs 1 and 2 meet this criterion. Next, use the following formula, and solve for x, the order of reaction with respect to A.

$$\frac{\text{Rate}_{Run\ 1}}{\text{Rate}_{Run\ 2}} = \left(\frac{[A]_{Run\ 1}}{[A]_{Run\ 2}}\right)^x$$

$$\frac{1.20 \times 10^{-2}}{1.20 \times 10^{-2}} = \left(\frac{0.100\ M}{0.200\ M}\right)^x$$

Solving for x, the order with respect to A is 0.

It is important to note that a zero order reaction does not have a rate of zero; it means the order is independent of the reaction concentration.

Runs 1 and 3 (the concentration of B changes and the concentrations of A and C remain constant) can be used to determine that the order with respect to B.

$$\frac{Rate_{Run\ 1}}{Rate_{Run\ 3}} = \left(\frac{[B]_{Run\ 1}}{[B]_{Run\ 3}}\right)^{y}$$

$$\frac{1.20 \times 10^{-2}}{4.80 \times 10^{-2}} = \left(\frac{0.200\ M}{0.400\ M}\right)^{y}$$

Solving for y, the order with respect to B is 2.

Runs 1 and 4 (the concentration of C changes and the concentrations of A and B remain constant) can be used to determine that the order with respect to C.

$$\frac{Rate_{Run\ 1}}{Rate_{Run\ 4}} = \left(\frac{[C]_{Run\ 1}}{[C]_{Run\ 4}}\right)^{z}$$

$$\frac{1.20 \times 10^{-2}}{2.40 \times 10^{-2}} = \left(\frac{0.200\ M}{0.400\ M}\right)^{z}$$

Solving for z, the order with respect to C is 1.

The overall rate law becomes

$$Rate = k[A]^0[B]^2[C]^1, \text{ or } Rate = k[B]^2[C]$$

The overall order of this reaction is determined by adding all of the orders; in this case, the overall order is 3 $(0 + 1 + 2)$.

The rate constant, k, with its appropriate units, can then be determined using the data from any one run by substituting the data from that run into the rate equation. The first run is used in this example:

$$1.2 \times 10^{-2}\ M/s = k[0.100\ M]^0[0.200\ M]^2[0.200\ M]$$

The units for k are determined algebraically $(Ms^{-1} = kM^3$; so $k = M^{-2}s^{-1})$, and k is determined to be $1.50\ M^{-2}s^{-1}$.

The overall rate law expression becomes:

$$Rate = 1.50\ M^{-2}s^{-1}[B]^2[C]$$

EXPERIMENT 12   THE RATE OF CHEMICAL REACTIONS

<u>Experimental Discussion</u>
The initial rate method will be used in this experiment. Solutions of known concentrations of I^-, $S_2O_8^{2-}$, and $S_2O_3^{2-}$ will be mixed together with a starch indicator. The time elapsed, Δt, from the start of mixing to the appearance of the dark color of the starch indicator will be measured. The two chemical reactions taking place are as follows:

$$2\,I^- \;+\; S_2O_8^{2-} \;\rightarrow\; I_2 \;+\; 2\,SO_4^{2-} \quad \text{(reaction equation)}$$

$$I_2 \;+\; 2\,S_2O_3^{2-} \;\rightarrow\; 2\,I^- + S_4O_6^{2-} \quad \text{(indicator reaction equation)}$$

In the first reaction equation, I_2 is produced. The I_2 will then react with the $S_2O_3^{2-}$ present according to the indicator reaction equation in a 1:2 stoichiometric ratio. As long as any $S_2O_3^{2-}$ is present, no free I_2 will be in solution. When the $S_2O_3^{2-}$ in solution is used up, the presence of I_2 will be detected by a starch indicator. (When I_2 and the iodide ion, I^-, exist in a solution where starch is present, a characteristic blue-black color will occur. The reaction is $I_2 \;+\; I^- \rightarrow\; I_3^-$. The triiodide ion slips into the coil of the starch causing the blue-black color.)

The rate of reaction will be measured by the increase in concentration of the product I_2.

$$\text{Rate} = \frac{\Delta[I_2]}{\Delta t}$$

As long as the concentrations of I^- and $S_2O_8^{2-}$ are much greater than the concentration of $S_2O_3^{2-}$, the conditions for the initial rate method will be satisfied and the rate equation will be of the form

$$\text{Rate}_{(initial)} = k[I^-]_0^x [S_2O_8^{2-}]_0^y$$

<u>Chemicals</u>

Potassium iodide	KI
Potassium persulfate	$K_2S_2O_8$ (or $(NH_4)_2S_2O_8$, Ammonium peroxydisulfate)
Sodium thiosulfate	$Na_2S_2O_3$
Potassium chloride	KCl
Sodium sulfate	Na_2SO_4
Iodine	I_2 (formed)
Starch	

<u>Waste</u>
    All waste can be disposed down the drain.

<u>Procedure</u>
    N.B.:   Rinse the beaker and Erlenmeyer flasks that are to contain the reaction mixtures with deionized water before each run and shake out any excess. It is not necessary to completely dry them before each run.

Table 12.2   Solutions for Kinetic Runs

Run	Flask A	Flask B	250 mL beaker
1	20.00 mL 0.200 M KI	20.00 mL 0.100 M $K_2S_2O_8$	10.00 mL 0.00500 M $Na_2S_2O_3$ 5 drops starch
2	20.00 mL 0.200 M KI	10.00 mL 0.100 M $K_2S_2O_8$ 10.00 mL 0.100 M Na_2SO_4	10.00 mL 0.00500 M $Na_2S_2O_3$ 5 drops starch
3	20.00 mL 0.200 M KI	15.00 mL 0.100 M $K_2S_2O_8$ 5.00 mL 0.100 M Na_2SO_4	10.00 mL 0.00500 M $Na_2S_2O_3$ 5 drops starch
4	10.00 mL 0.200 M KI 10.00 mL 0.200 KCl	20.00 mL 0.100 M $K_2S_2O_8$	10.00 mL 0.00500 M $Na_2S_2O_3$ 5 drops starch
5	15.00 mL 0.200 M KI 5.00 mL 0.200 M KCl	20.00 mL 0.100 M $K_2S_2O_8$	10.00 mL 0.00500 M $Na_2S_2O_3$ 5 drops starch

For each of the above, measure the KI and the $K_2S_2O_8$ precisely from the burets. The KCl and Na_2SO_4 solutions are added only to keep the total ionic concentration in each mixture the same. This is necessary to avoid secondary effects.

1. Calculate the initial concentrations of I^-, $S_2O_8^{2-}$, and $S_2O_3^{2-}$ in each run using the data in Table 12.2 and the dilution formula $M_1V_1 = M_2V_2$, is as follows:

$$[I^-]_0 = \frac{(0.200 \text{ M})(20.00 \text{ mL})}{(50.00 \text{ mL})} = 0.0800 \text{ M}$$

$$\left[S_2O_8^{2-}\right]_0 = \frac{(0.100 \text{ M})(20.00 \text{ mL})}{(50.00 \text{ mL})} = 0.0400 \text{ M}$$

$$\left[S_2O_3^{2-}\right]_0 = \frac{(0.00500 \text{ M})(10.00 \text{ mL})}{(50.00 \text{ mL})} = 0.00100 \text{ M}$$

2. Calculate $\Delta[I_2]$. From the indicator reaction equation, the concentration of $S_2O_3^{2-}$ is twice as great as the concentration of I_2 produced in the time Δt. <u>Therefore, the concentration of I_2 produced in time Δt will be one-half the concentration of $S_2O_3^{2-}$ initially present in the reaction solution.</u>

$$[S_2O_3^{2-}] = 1.00 \times 10^{-3} \text{ M; therefore } \Delta[I_2] = 5.00 \times 10^{-4} \text{ M}$$

3. Normalize and then fill four burets: one with 0.200 M KI, one with 0.200 M KCl, one with 0.100 M $K_2S_2O_8$ and the remaining one with 0.100 M Na_2SO_4.

4. Prepare the solutions for Run 1 as indicated in Table 12.2.

5. Add 10.00 mL of 0.00500 M $Na_2S_2O_3$ by pipet and 5 drops of starch indicator to the 250 mL beaker.

6. Quickly pour the two solutions in the flasks into the beaker and swirl the solution gently during the reaction to mix. Time the reaction from the moment of mixing to the instant the solution turns color, ranging from a light brown to a blue-black. One partner should gently shake or swirl the beaker to mix the solutions, and the other should pour and read the clock.

7. For Runs 2 to 5, prepare the solutions listed in Table 12.2. Repeat steps 5 and 6 for each run.

LABORATORY REPORT SHEET

Name_______________________________________   Date___________________

Partner_____________________________________   Temperature______°C

Run	$[I^-]_0$	$[S_2O_8^{2-}]_0$	$[S_2O_3^{2-}]_0$	$\Delta[I_2]$	Δt	Rate $(\Delta[I_2]/\Delta t)$	k
Units:							
1							
2							
3							
4							
5							

1. Determine the order with respect to $[I^-]$.

2. Determine the order with respect to $[S_2O_8^{2-}]$.

3. Determine the overall order of the reaction.

4. Calculate the average k (with units).

5. Write the overall rate law expression with the k value specific to this reaction.

Error Analysis

1. Assuming the identity and the concentrations of the reactants remain the same in a set of experiments,
    a. What is the only physical parameter that can change the rate constant?

    b. How exactly will the rate be affected if this parameter is increased?

2. In a set of reactions, the concentration of the reactants is purposely changed in each run to determine the rate law.
    a. If the time of the reaction increases from one run to the next, how would this affect the _experimentally measured_ rate?

    b. How would this affect the rate constant?

    c. Circle the experimental errors that could falsely increase the time of a reaction if the concentrations of the reactants remain the same.

    Stopping the time too late   Not adding the flask contents at the same time

    Not swirling adequately   Adding too little $S_2O_3^{2-}$, but recording 10.00 mL

    Warming up the contents of the flasks   Stopping the time too early

3. Assuming all volumes remain the same, if the 0.200 M KI used in this experiment was replaced with 0.100 M KI,
    a. How would this affect the rate of the reaction?

    b. How would this affect the rate constant?

4. Complete the following two sentences for the following rate law: Rate $= k\,[A]^2[B]^0[C]^1$

    This reaction is ______________ order with respect to A, ______________ order

    with respect to B, and ______________ order with respect to C. The overall

    order is ______________ , and the units of the rate constant are ______________ .

5. Give a practical example of a reaction whose rate is measured.

HOMEWORK EXERCISES

1. Four solutions listed below are made by mixing the reagents as indicated at 25 °C. In each case the total volume of the solutions is 100.00 mL. **Note: $\Delta[Se] = [H_2SeO_3]$**

$$H_2SeO_3 + 6\,I^- + 4\,H^+ \rightarrow Se + 2\,I_3^- + 3\,H_2O$$

Solution 1
25.00 mL of 0.200 M H_2SeO_3
25.00 mL of 1.20 M KI
25.00 mL of 0.800 M HCl
25.00 mL KCl

Solution 2
50.00 mL of 0.200 M H_2SeO_3
25.00 mL of 1.20 M KI
25.00 mL of 0.800 M HCl

Solution 3
25.00 mL of 0.200 M H_2SeO_3
50.00 mL of 1.20 M KI
25.00 mL of 0.800 M HCl

Solution 4
25.00 mL of 0.200 M H_2SeO_3
25.00 mL of 1.20 M KI
50.00 mL of 0.800 M HCl

Run	$[H_2SeO_3]_0$	$[I^-]_0$	$[H^+]_0$	$\Delta[Se]$	Δt	Rate ($\Delta[Se]/\Delta t$)	k
Units					sec		
1					300		
2					300		
3					37.5		
4					75		

What is the order of the reaction with respect to H_2SeO_3?

What is the order of the reaction with respect to I^-?

What is the order of the reaction with respect to H^+?

What is the overall order of reaction? __________

Write the expression for the rate law ____

Calculate the k values for each run.

Write the rate law with the average k value and units ________________________________

Continued

Use the concentration-time graphical method (on graph paper) for problems 2 and 3. This can be done on the computer.

2. $2 N_2O_5 \rightarrow 2 N_2O_4 + O_2$ (at 25 °C)

t	$[N_2O_5]$	$\ln[N_2O_5]$	$1/[N_2O_5]$
25 sec	0.800 M		
50 sec	0.400 M		
75 sec	0.200 M		
100 sec	0.100 M		

a. Graphically, determine the order of the reaction with respect to N_2O_5. _________

b. Write the rate law for this reaction.

c. Calculate k for this reaction.

3. $2 NOCl \rightarrow 2 NO + Cl_2$ (at 25 °C)

t	[NOCl]	$\ln[NOCl]$	$1/[NOCl]$
56.25 sec	0.800 M		
62.50 sec	0.400 M		
75.00 sec	0.200 M		
100.0 sec	0.100 M		

a. Graphically, determine the order of the reaction with respect to NOCl. _______

b. Write the rate law for this reaction.

c. Calculate k for this reaction.

EXPERIMENT 13
QUANTITATIVE ANALYSIS OF AN ALLOY

An alloy containing several metals will be dissolved and the resulting solution will be analyzed for the percent of copper and silver. An instrument called a spectrophotometer (commonly referred to as a spectrometer) will be used for the copper determination and titration will be used for the silver analysis.

Key Chemical Reactions:

$$4\ HNO_{3\ (aq)} + 3\ Ag_{(s)} \rightarrow 3\ AgNO_{3\ (aq)} + 2\ H_2O_{(l)} + NO_{(g)}$$

$$4\ HNO_{3\ (aq)} + Cu_{(s)} \rightarrow Cu(NO_3)_{2\ (aq)} + 2\ NO_{2\ (g)} + 2\ H_2O_{(l)}$$

$$Ag^+_{\ (aq)} + SCN^-_{\ (aq)} \rightarrow AgSCN_{(s)}$$

$$Fe^{3+}_{\ (aq)} + SCN^-_{\ (aq)} \rightarrow Fe(SCN)^{2+}_{\ (aq)}$$

Key Mathematical Equations:

$$A = \varepsilon b C$$

$$L\ KSCN \times M\ KSCN \times \frac{1\ mole\ Ag}{1\ mole\ KSCN} \times \frac{107.87\ g\ Ag}{1\ mole\ Ag} = g\ Ag$$

$$5 \times g\ Ag\ in\ aliquot = g\ Ag\ in\ 50.0\ mL$$

$$\frac{g\ Ag\ in\ 50.0\ mL}{mass\ of\ alloy\ sample} \times 100 = \%\ Ag\ in\ alloy$$

<u>Discussion</u>

In a spectrometer, a beam of light at a selected wavelength is passed through a solution and strikes a photoelectric detector. If the wavelength of the light is the same as that absorbed by a species in the solution, some light will be absorbed and the intensity of the light emerging from the solution will be lower than it was before it entered the solution. By comparing the intensity of light before and after it passes through the solution, the concentration of the absorbing species can be calculated. The schematic of a spectrometer is shown in Figure 13.1.

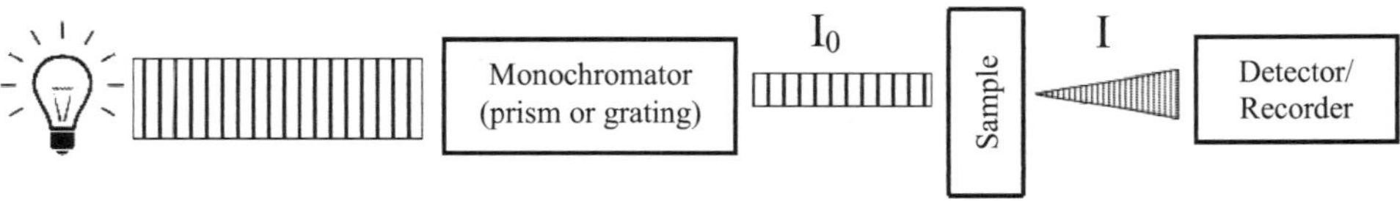

Figure 13.1. Schematic diagram of a spectrophotometer

In Figure 13.1, I_0 is the symbol for the intensity of light incident on the sample, which in this case is a cell containing a copper solution. I is the intensity of light transmitted through the solution. By definition, A, the absorbance of the sample, is equal to the log of the quotient I_0 / I.

$$A = \log \frac{I_0}{I}$$

For dilute solutions, the absorbance is directly proportional to the concentration of the absorbing species present in the solution. Therefore, a plot of the absorbance, A, at a particular wavelength versus concentration will be linear. This relationship is commonly known as Beer's Law, where A is the absorbance, b is the cell path length (usually 1 cm), ε is the molar extinction coefficient or molar absorptivity, (L/mol-cm), a constant for a species, and C is the concentration (mole/L).

$$A = \varepsilon\, b\, C$$

The instrument is first calibrated by setting the absorbance of a cuvette filled with the pure solvent (usually DI water) equal to zero. Then the cuvette is filled with the solution, and its absorbance is read on the spectrometer.

The first step in this experiment is to dissolve the alloy, which is done by the addition of nitric acid, as shown in the following equations:

$$4\ HNO_{3\ (aq)} + 3\ Ag_{(s)} \rightarrow 3\ AgNO_{3\ (aq)} + 2\ H_2O_{(l)} + NO_{(g)}$$

$$4\ HNO_{3\ (aq)} + Cu_{(s)} \rightarrow Cu(NO_3)_{2\ (aq)} + 2\ NO_{2\ (g)} + 2\ H_2O_{(l)}$$

Both the silver nitrate and copper nitrate dissociate in aqueous solution giving the Ag^+ and Cu^{2+} ions. In the dissolved alloy sample, the only metal present whose ions are colored in aqueous solution is copper. The Cu^{2+} is blue in color, and the wavelength that will be used for the copper determination by the spectrometer is 800 nm.

The silver content of the alloy will be determined by the Volhard method, which involves titration of the alloy solution (Ag^+ ions) with a standard solution of KSCN using the Fe^{3+} ion as an indicator. AgSCN is insoluble in water, so the addition of KSCN causes precipitation of the Ag^+ in solution:

$$Ag^+_{\ (aq)} + SCN^-_{\ (aq)} \rightarrow AgSCN_{(s)}\ \text{(white precipitate)}$$

When all the silver ion has been effectively precipitated from the solution, the next increment of KSCN will cause the appearance of the $Fe(SCN)^{2+}$ complex, which has a brown-red color:

$$Fe^{3+}_{\ (aq)} + SCN^-_{(aq)} \rightarrow Fe(SCN)^{2+}_{\ (aq)}\ \text{(brown-red solution)}$$

Chemicals

Silver	$Ag_{(s)}$
Copper	$Cu_{(s)}$
Nitric acid, 8 M	$HNO_3{}_{(aq)}$
Sodium hydroxide, 6 M	$NaOH_{(aq)}$
Potassium thiocyanate	$KSCN_{(aq)}$
Ferric alum indicator	$Fe(NH_4)(SO_4)_2 \cdot 12\ H_2O$

Waste

All waste containing copper or silver (including normalizing solutions) should be collected and disposed of in the waste container. Excess potassium thiocyanate solution and gas trap waste can be disposed down the drain.

Procedure

Dissolution of the Sample

1. Add two boiling chips that have been rinsed with DI water to a large test tube. Add 50.0 mL of DI water, carefully measured, to the large test tube and mark the level with tape or grease pencil. (Record the volume to 3 significant figures, 50.0 mL.)

2. Carefully pour out the water leaving the two boiling chips in the test tube. Clamp the test tube to the ring stand at the <u>top of the test tube</u>.

3. Weigh approximately 0.5 g of the alloy sample and transfer it to the large test tube.

4. Assemble the same trap apparatus that used in the Copper Cycle experiment (see Figure 2.1), with the gas trap (Erlenmeyer flask) containing about 150 mL of tap water plus about 10 mL of 6 M NaOH solution. The oxides of nitrogen evolved during the solution of the sample are very toxic and must not be allowed to escape into the room; the gas trap neutralizes them.

5. <u>With the aspirator on</u>, add two disposable pipetfuls of 8 M nitric acid to the alloy sample and slide the glass hood over the test tube as shown in Figure 2.1. Ensure no sample remains on the side of the test tube: target the addition of nitric acid to "wash" any sample down to the bottom, then use a very small of water if still necessary. <u>Nitric acid is a highly corrosive liquid</u>. If any is spilled, remove it using copious quantities of water.

6. Begin heating the reaction tube gently by holding the Bunsen burner under the test tube for 10-15 seconds at a time, then removing it, taking care not to let any of the solution "bump" over into the gas trap. While heating, continue to apply suction from the gas trap at an intermediate rate. If any sample remains on the walls, wash it down with a small amount of water. After the sample has completely dissolved, boil gently for about one minute to expel the oxides of nitrogen (NO_x is indicated by a solution that is green and by the presence of brown gas on the inside of the test tube.) The resulting NO_x-free solution should be blue, and no brown gas should be present on the sides of the test tube. Shut off the aspirator when done.

7. After the solid has been dissolved and the oxides of nitrogen have been expelled, dilute the alloy solution to the 50.0 mL mark with deionized water.

8. Stopper the test tube and mix the solution by inverting the test tube a few times. If the solution is clear, proceed with the determination of copper and silver.

<u>Determination of Copper</u>

The accuracy of the spectrophotometric procedure for copper can be drastically reduced by adhering to the following:

- Use the same cuvette for calibration and analyzing of the samples, since cuvettes can differs appreciably in their light transmitting properties.
- Normalize the cuvette before taking each absorbance reading.
- Hold the cuvette only by the top, and wipe off the faces of the cuvette before inserting into the spectrometer. Fingerprints and dirt can refract light and change absorbance values.

1. Set up the computer and spectrometer and start the spectrometer program.

2. Calibrate the spectrometer with DI water according to the procedure outlined in the given directions.

3. Obtain 10-15 mL of each copper calibration solution. Normalize the cuvette with the first known concentration of a copper solution and then fill the cuvette to about a quarter inch from the top using a disposable pipet. Wipe the sides of the cuvette with a Kimwipe and then determine the absorbance of the solution at 800 nm (absorbance is the red number on the bottom tool bar). Dispose of the solution in the waste container. Repeat with the other known concentrations of copper and DI water.

4. Normalize a cuvette with the alloy solution and then fill the cuvette to about a quarter inch from the top using a disposable pipet. Determine the absorbance of the solution at 800 nm. Dispose of the solution in the waste container.

5. Construct a graph of absorbance versus concentration of copper in grams per 50.0 mL of solution using the known copper concentrations and their absorbances. Determine the equation of the line.
    - Minimize the spectrometry program and start Excel.
    - Enter the g Cu/50 mL values in column 1 and the absorbance values in column 2.
    - Highlight the data, click the Insert tab, and select the Scatter graph.
    - Label both the x and y axes of the graph, and give the graph a title.
    - Right click on any point, click Add Trendline, and select Display Equation on Chart.
    - Right click in the graph, select Move Chart, and select New Sheet.
    - Print a copy for each lab partner.
    - Click close and <u>DO NOT SAVE</u> the Excel spreadsheet.

5. Using the equation and the absorbance reading of the unknown, determine the mass of copper/50.0 mL solution and calculate the percent copper in the alloy sample.

6. Turn the spectrometer's strobe off and <u>shut down the computer through the start menu</u>.

<u>Determination of Silver</u>

1. Normalize a pipet and then pipet a 10.00 mL aliquot of the alloy solution into a clean 125 mL Erlenmeyer flask. Add about 25 mL of deionized water to bring the solution up to a convenient level.

2. Add about 15 drops of ferric alum indicator to the solution and then titrate with standardized KSCN solution to the endpoint. The endpoint will occur when the solution is a pale brown-orange. The solution will also have a chalky appearance due to the precipitation of AgSCN. This precipitate should not be confused with the endpoint of the titration, which is signaled by a color change. Calculate the mass of the silver in the aliquot and then the mass and percent silver in the solution.

Mass of silver in the aliquot:

$$\text{L KSCN x M KSCN x } \frac{1 \text{ mole Ag}}{1 \text{ mole KSCN}} \text{ x } \frac{107.87 \text{ g Ag}}{1 \text{ mole Ag}} = \text{g Ag}$$

Mass of silver in the sample:

$$5 \text{ x g Ag in aliquot} = \text{g Ag in 50.0 mL}$$

% Ag in sample:

$$\frac{\text{g Ag in 50.0 mL}}{\text{mass of alloy sample}} \text{ x } 100 = \text{\% Ag in alloy}$$

3. Drain the remaining KSCN in the buret to the drain. Fill the buret with DI water, drain, and return the buret to the buret clamp, upside down with the stopcock open.

4. Transfer all hazardous waste collected to the hazardous waste carboy.

LABORATORY REPORT SHEET

RECORD ALL VALUES
WITH PROPER UNITS AND
SIGNIFICANT FIGURES!

Name_________________________________   Date___________________________

Unknown Number __________________   Mass of alloy sample __________________

Copper

Calibration Data	Absorbance
0.00 g Cu/50.0 mL solution (DI water)	
0.10 g Cu/50.0 mL solution	
0.15 g Cu/50.0 mL solution	
0.20 g Cu/50.0 mL solution	

Equation of best-fit line: ____

Absorbance of unknown solution ___________________

Mass of Cu in solution (using the above equation), g/50.0 mL ___________________

% Cu in sample:

Silver

Initial buret reading ________________   Molarity of KSCN solution ________________

Final buret reading ________________

Volume of KSCN delivered ________________

Mass of Ag present in aliquot:

Mass of Ag present in 50.0 mL sample:

% Ag in sample:

Error Analysis

<u>Example (for Questions 1-3)</u>
If the copper calibration curve was determined at 800 nm, but the sample was measured at 690 nm, what species would this impact, what would be the effect on the quantity of the parameter measured, and exactly how would the final results appear to be affected?

> *The species <u>copper</u> would be impacted. The <u>absorbance of the sample would be lower</u> at 690 nm than 800 nm. This would result in a <u>lower g Cu/50.0 mL solution value</u> read from the graph. The <u>% Cu would appear lower</u> than the actual value.*

1. If there were scratches or fingerprints on the cuvette, what species would this impact, what would be the effect on the quantity of the parameter measured, and exactly how would the final results appear to be affected?

2. If the solution was overtitrated with potassium thiocyanate, what species would this impact, what would be the effect on the quantity of the parameter measured, and exactly how would the final results appear to be affected? (Be sure to define overtitration in the answer.)

3. If the alloy sample in the test tube was accidentally diluted to a volume above the 50.0 mL mark, but a total volume of 50.0 mL was assumed, what species would this impact, what would be the effect on the quantity of the parameter measured, and exactly how would the final results appear to be affected?

4. Beer's Law states that the absorption is equal to the product of the molar absorptivity, ε, (a constant for a species) times the path length of a cell (b) times the concentration (C) of a solution: $A = \varepsilon\, b\, C$.

    If the cell path length is 1.0 cm, such as the cell used in this experiment, what are the units of molar absorptivity? Show work.

Name ________________________________

HOMEWORK EXERCISES

Consider the plot drawn below for the questions 1 and 2.

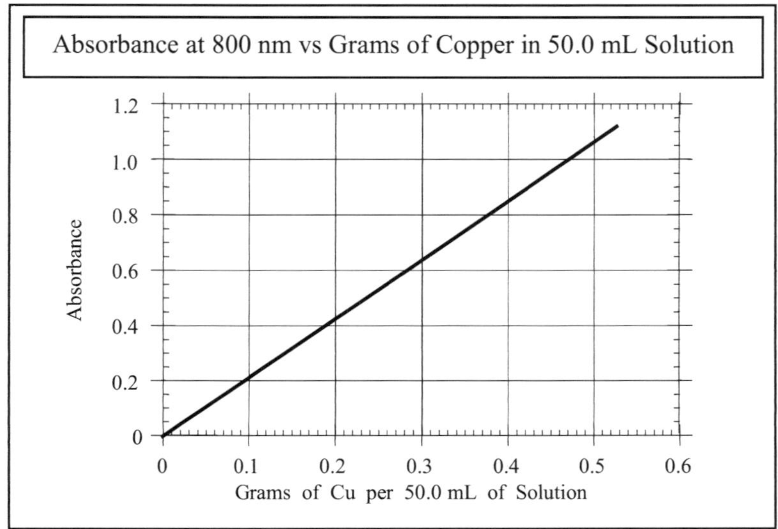

1. When 0.7587 g of an alloy is dissolved in a total volume of 50.0 mL, the absorbance of the solution at 800 nm, due to copper, is found to be 0.675. A 10.00 mL aliquot of the solution is titrated with 0.0392 M KSCN solution and 8.25 mL of the KSCN solution is required to reach the end point. Calculate:

   a. the grams of copper present in the alloy sample

   b. the percent copper in the alloy

   c. the grams of silver in the aliquot

   d. the grams of silver in the alloy

   e. the percent silver in the alloy

Continued on back

2. A 0.4682 g sample of a metal is dissolved in a total volume of 50.0 mL.
    a. If the metal was pure silver, how many mL of a 0.1215 M KSCN solution would be
        required to react with a 5.00 mL aliquot of the alloy solution?

    b. If the metal was pure copper, what would be the absorbance of the solution?

3. a. Write balanced molecular equation for the reaction of silver with concentrated nitric
        acid. Include states.

    b. Write balanced molecular equation for the reaction of copper with concentrated nitric
        acid. Include states.

    c. What toxic gases are formed in these reactions? ___________________________

    d. What component neutralizes the toxic gas? ___________________________

4. a. Write the molecular, ionic, and net ionic equations for the reaction of silver nitrate
        with potassium thiocyanate. Include states.

Molecular:

Ionic:

Net ionic:

    b. What indicator is used in this titration? ___________________________

    c. Explain in detail exactly how this indicator works in this reaction.

EXPERIMENT 14
LE CHÂTELIER'S PRINCIPLE

The relationship between reactants and products when a stress is placed on a system and how each stress affects the equilibrium position will be investigated in terms of Le Châtelier's Principle.

Key Chemical Reactions:

$$2\,CrO_4^{2-} \;+\; 2\,H^+ \;\rightleftharpoons\; Cr_2O_7^{2-} \;+\; H_2O$$

$$HIn \;+\; OH^- \;\rightleftharpoons\; In^- \;+\; H_2O$$

$$Fe^{3+} \;+\; SCN^- \;\rightleftharpoons\; Fe(SCN)^{2+}$$

$$CoCl_4^{2-} \;+\; 6\,H_2O \;\rightleftharpoons\; 4\,Cl^- \;+\; Co(H_2O)_6^{2+} \;+\; heat$$

$$BaCrO_{4\,(s)} \;\rightleftharpoons\; Ba^{2+}_{\;(aq)} \;+\; CrO_4^{2-}{}_{(aq)}$$

Key Mathematical Equations: None

Discussion

Le Châtelier's Principle states that when a system at equilibrium is subjected to an external stress, the system will shift in a direction to counteract the stress and achieve a new equilibrium state. External stresses that could affect a system include changes in concentration, volume, pressure, and temperature, or the addition of a catalyst. Each of these will be examined in this discussion of their effect on the system and equilibrium constant, K_c.

In order to understand Le Châtelier's Principle, consider the following reaction:

$$PCl_{3\,(g)} \;+\; Cl_{2\,(g)} \;\rightleftharpoons\; PCl_{5\,(g)} \qquad\qquad \Delta H° = -\,88\ kJ$$

Since this reaction is exothermic (heat is being released), it can be rewritten as follows:

$$PCl_{3\,(g)} \;+\; Cl_{2\,(g)} \;\rightleftharpoons\; PCl_{5\,(g)} \;+\; 88\ kJ$$

Changes in Concentration

If the system is subjected to changes in concentration, the system will adjust to counteract the stress to achieve a new equilibrium position. In the above reaction, if more PCl_3 is added, the "stress" is more PCl_3, and the system is no longer in equilibrium. In order to achieve equilibrium, the system will move in a direction to counteract the stress and remove PCl_3. To remove the PCl_3, Cl_2 will react with it, and more PCl_5 will be formed.

Therefore, the reaction shifts in the forward direction ("to the right"). The PCl_3 and Cl_2 concentrations will decrease and the PCl_5 concentration will increase. Similarly, if more product is added, such as PCl_5, the "stress" is more PCl_5, and PCl_5 would need to be removed. The reaction would then shift in the reverse direction ("to the left") and the concentrations of PCl_3 and Cl_2 would increase.

A corollary to this would be the removal of a reactant or product. If PCl_3 was removed, the stress on equilibrium would be too little PCl_3. To counteract this stress, PCl_5 would form PCl_3 and Cl_2, and the reaction would shift to the left. If PCl_5 was removed, the stress on equilibrium would be too little product, and the reaction would shift to the right (decreasing the amounts of PCl_3 and Cl_2) to form more product.

Both of these examples can alter the concentration of the reacting mixture, but they do not change K_c. It should also be noted that the concentration of a solid, like its density, is an intensive property and does not depend on how much of the substance is present. Therefore, solids are not included in a K_c equation since they do not affect the equilibrium. Pure liquids in heterogeneous equilibria also do not appear in equilibrium expressions.

Changes in Volume and Changes in Pressure
Changes in volume and pressure usually do not affect the concentrations of reacting species in condensed phases because liquids and solids are generally incompressible. However, gases are affected by changes in volume and pressure, such as in a moveable piston. To understand these changes, the ideal gas equation is used.

$$PV = nRT$$

Rearranging, this equation becomes

$$P = \left(\frac{n}{V}\right)RT$$

where $\left(\frac{n}{V}\right)$ is the concentration of the gas, (M)

P and V are inversely proportional to each other, so that if the pressure is increased, the volume decreases.

There are three ways to change the pressure of a system: (1) change the volume, (2) add an inert gas, such as He or N_2, or (3) change the concentration of a reactant or product.

Again, consider the following equation.

$$PCl_{3\,(g)} + Cl_{2\,(g)} \rightleftharpoons PCl_{5\,(g)} + 88\ kJ$$

If the volume of the system is increased (or the pressure is decreased), the "stress" is more available space for gas molecules in the system. The system tries to produce more gas (make more moles of gas) to fill the space. The reaction shifts in the direction of the reaction that makes more <u>moles of gas</u>. In this case, there are two moles of gaseous reactants and one mole of gaseous products. The reaction would shift in the reverse direction, toward the side with more moles of gaseous species. Therefore, the concentrations of PCl_3 and Cl_2 would increase. Conversely, if the volume is decreased (or the pressure is increased), the reaction would shift in the direction of the side of fewer moles of gas. In this example, the reaction would shift in the forward direction, to the right, and more moles of PCl_5 would be formed.

When adding an inert gas such as N_2 or He to a system at constant volume, the pressure of the system increases. Consequently, the mole fraction of gaseous constituents decreases, but the partial pressure of each gas remains the same (mole fraction x total pressure). Therefore, the addition of an inert gas in a system of constant volume does not affect the equilibrium.

Changes in volume and pressure of a system with some gaseous constituents can alter the concentration of the reacting mixture, but they do not change K_c.

Changes in Temperature
Changes in temperature require knowing whether the reaction is exothermic or endothermic ($\Delta H°$). Again, consider the following equation, where $\Delta H° = -\,88$ kJ.

$$PCl_{3\,(g)} \; + \; Cl_{2\,(g)} \; \rightleftharpoons \; PCl_{5\,(g)} \; + \; 88 \text{ kJ}$$

If the temperature of the system is increased, the "stress" added is heat, and the system needs to remove heat in order to establish equilibrium. In this reaction, since the reaction is exothermic, heat is a product, and adding a product causes the reaction to shift in reverse or to the left. Therefore, the concentrations of PCl_3 and Cl_2 would increase. In other words, a temperature increase favors the endothermic direction of a reaction and a temperature decrease favors the exothermic direction of a reaction.

Changes in temperature alter the concentration of the reacting mixture and is the only stress that can change the equilibrium constant (K_c) of a reaction.

Changes in the Addition of a Catalyst
The addition of a catalyst does not affect the position of equilibrium, and there is no change in any of the equilibrium concentrations or K_c. The catalyst merely lowers the time needed to attain equilibrium.

In this experiment several equilibria and the effect of an applied stress on the system will be studied. The stresses that will be applied include adding and removing reactants and products, and temperature changes. The change in volume or pressure will not be studied in this experiment because these changes are difficult to measure in the General Chemistry laboratory.

Chemicals: (1 mL = 20 drops)
>Water
>Potassium chromate, K_2CrO_4 (0.1 M)
>Sulfuric acid, H_2SO_4 (6 M)
>Sodium hydroxide, NaOH (6 M)
>Bromocresol green indicator solution
>Phenolphthalein indicator solution
>Hydrochloric acid, HCl (6 M and 12 M)
>Iron (III) nitrate, $Fe(NO_3)_3$ (0.1 M)
>Potassium thiocyanate, KSCN (0.1 M)
>Silver nitrate, $AgNO_3$ (0.1 M)
>Cobalt (II) chloride, $CoCl_2$ (0.1 M)
>Barium chloride, $BaCl_2$ (0.1 M)

Note: Explain in terms of products and reactants ("adding/removing", etc.).
- In the "Stress on System" box, write either "add" or "remove" with the **ion** or parameter (like temperature) that is stressing the system.
- To determine if it is an "addition" or "removal" in the "(I)" column, look to see if the stressed ion appears in the balanced net ionic equation. If it does, it is an "addition". If it does not appear in the net ionic equation, determine if the stressed ion could react with a species in the net ionic equation; this would be a "removal".

Record observations on the lab sheet in the following form:

Step	Stress on System	Direction of Shift	Observations (O) and Interpretations (I)
A.3	*Add H^+*	*Right*	(O): *Solution turned orange*
	Q.A.1. Explain what is happening in step A.3 in terms of Le Châtelier's Principle.		(I): *Addition of a reactant shifts the equilibrium to the products.*

Waste
>The wastes from Procedure Parts A, C, D, and E should be collected in the waste beaker and then disposed of in the laboratory waste container. Wastes from Part B and any excess $Fe(SCN)^{2+}$ solution in Part C can be disposed down the drain. Each part details these instructions.

EXPERIMENT 14   LE CHÂTELIER'S PRINCIPLE

<u>Procedure:</u> (Test tubes used are size 13 x 100 mm)

A. *Chromate – Dichromate Equilibrium*

$$2\,CrO_4^{2-} \;+\; 2\,H^+ \;\rightleftharpoons\; Cr_2O_7^{2-} \;+\; H_2O$$

The effect on equilibrium can be observed due to color changes: the CrO_4^{2-} ion is yellow and the $Cr_2O_7^{2-}$ ion is orange.

1. Add approximately 3 mL (~1 inch) of 0.1M K_2CrO_4 to a test tube.

Adding a reactant:
2. Add 3 – 5 drops of 6 M H_2SO_4.
3. Mix, record any observations, and answer data question Q.A.1.

Removing a reactant:
4. Add 10 drops of 6 M NaOH.
5. Mix, record any observations, and answer data question Q.A.2.
6. Add 5 drops of 6 M H_2SO_4.
7. Mix, record any observations, and answer data question Q.A.3.
8. Dispose of the waste from Part A in the appropriate waste container. Do not pour this down the drain!

B. *Weak Acid – Weak Base Equilibrium*

The effect on equilibrium can be observed due to color changes: the indicator bromocresol green is yellow in its acidic form and blue in its basic form, and the indicator phenolphthalein is colorless in its acidic form and pink in its basic form. They are also weak acid – weak base equilibrium systems.

$$HIn \;+\; OH^- \;\rightleftharpoons\; In^- \;+\; H_2O$$

where   HIn is the acid form of the indicator, and
                   In^- is the basic form of the indicator

<u>Bromocresol Green Indicator</u>
1. Add approximately 3 mL of DI water to a test tube.
2. Add one drop of bromocresol green indicator solution.
3. Mix, record any observations, and answer data question Q.B.1.
4. Add 2 drops of 6 M NaOH.
5. Mix, record any observations, and answer data question Q.B.2.
6. Add 4 drops of 6 M HCl.
7. Mix, record any observations, and answer data question Q.B.3.
8. Dispose of the waste in the sink.

<u>Phenolphthalein Indicator</u>
9. Add approximately 3 mL of DI water to a clean test tube.
10. Add one drop of phenolphthalein indicator solution.
11. Mix, record any observations, and answer data question Q.B.4.
12. Add 2 drops of 6 M NaOH.
13. Mix, record any observations, and answer data question Q.B.5.
14. Add 4 drops of 6 M HCl.
15. Mix, record any observations, and answer data question Q.B.6.
16. Dispose of the waste in the sink.

C. *Complex Ion Equilibrium: Fe(SCN)$^{2+}$ ion*

$$Fe^{3+} \; + \; SCN^- \; \rightleftharpoons \; Fe(SCN)^{2+}$$

The effect on equilibrium can be observed due to color changes: the Fe^{3+} ion is yellow and the $Fe(SCN)^{2+}$ ion is red.

1. Add 1 pipetful of 0.1 M $Fe(NO_3)_3$ and 1 pipetful of 0.1 M KSCN to a 100 mL beaker. Add 50 mL of DI water to the beaker.
2. Pour about 5 mL of the stock solution into five different test tubes.
3. Test tube #1 is used as a control for comparison. Record any observations, and answer data question Q.C.1.
4. To test tube #2, add 1 mL of 0.1 M $Fe(NO_3)_3$. Mix, record any observations, and answer data question Q.C.2.
5. To test tube #3, add 1 mL of 0.1 M KSCN. Mix, record any observations, and answer data question Q.C.3.
6. To test tube #4, add 6 drops of NaOH. Mix, record any observations, and answer data question Q.C.4. (OH^- reacts to form an insoluble precipitate. This is not an equilibrium reaction, so the reaction is a single arrow.)
7. To test tube #5, add 4 drops of 0.1 M $AgNO_3$. Mix, record any observations, and answer data question Q.C.5. (Ag^+ reacts to form an insoluble precipitate. This is not an equilibrium reaction, so the reaction is a single arrow.)
8. Dispose of the waste in the test tubes in the appropriate waste container. The stock solution in the 250 mL beaker can be disposed down the drain.

D. *Complex Ion Equilibrium: CoCl$_4$$^{2-}$ and Co(H$_2$O)$_6$$^{2+}$ ions*

$$CoCl_4^{2-} \; + \; 6\,H_2O \; \rightleftharpoons \; 4\,Cl^- \; + \; Co(H_2O)_6^{2+} \; + \; heat$$

The effect on equilibrium can be observed due to color changes: the $CoCl_4^{2-}$ ion (the "chloro" complex) is purple-blue and the $Co(H_2O)_6^{2+}$ (the "aqua" complex) ion is pink. The conversion from one form to the other involves an energy change, and therefore the equilibrium is temperature dependent.

1. Make a hot water bath by placing a 250 mL beaker half full of water on a hot plate. Heat the water so that the temperature is warm, but not boiling. Proceed to step 2 while the water is heating.
2. Add 5 mL (~ 2 inches) of 0.1 M $CoCl_2$ prepared in methanol to a dry test tube.
3. Add (dropwise) a minimal amount of DI water to the blue $CoCl_2$ complex to just change the color to that of the pink aqua complex. DO NOT ADD TOO MUCH WATER.
4. Divide the pink solution equally into three test tubes.
5. To test tube #1, add 12 M HCl dropwise until a color change is observed. Mix, record any observations, and answer data questions Q.D.1 – Q.D.3.
6. Place test tube #2 in the hot water bath. Record any observations, and answer data question Q.D.4.
7. Place test tube #3 in an ice bath. Record any observations, and answer data questions Q.D.5 and Q.D.6.
8. Dispose of the waste in the appropriate waste container.

E. *Saturated Solution Equilibrium: Barium chromate*

$$BaCrO_{4\ (s)} \rightleftharpoons Ba^{2+}_{\ (aq)} + CrO_4^{2-}_{\ (aq)}$$

The effect on equilibrium can be observed due to the appearance of a precipitate.

1. Add 3 mL of 0.1 M $BaCl_2$ to a test tube and then add 5 drops of K_2CrO_4. Mix with a stirring rod. This yields solid $BaCrO_4$.
2. Centrifuge for 20-30 seconds. When centrifuging, place a test tube with an equal amount of water as the sample across from the sample so that the centrifuge is balanced.
3. Decant the liquid into the waste beaker. Wash the precipitate with 3 mL of DI water. Decant the wash liquid, discarding the decanted liquid in the waste beaker.
4. Add 5 mL of DI water to the precipitate and shake or stir for three minutes to establish the equilibrium position. Record any observations.
5. Add 8 drops of 6 M HCl and shake. Record any observations, and answer data question Q.E.1. In this reaction, the hydrogen ion reacts with the chromate ion to form chromic acid, which is a weak electrolyte/weak acid.
6. Dispose of the waste in the appropriate waste container.

EXPERIMENT 14 LE CHÂTELIER'S PRINCIPLE

A. $2\,CrO_4^{2-} + 2\,H^+ \rightleftharpoons Cr_2O_7^{2-} + H_2O$

Ion
Color: __________   __________

Step	Stress on System	Direction of Shift	Observations (O), Interpretations (I), and Reactions (R)
A.3			(O):
	Q.A.1. Explain what is happening in step A.3 in terms of Le Châtelier's Principle.		(I):
A.5			(O):
	Q.A.2. (a) Explain what the addition of NaOH in this step does to the reaction in terms of Le Châtelier's Principle. (b) Support this explanation by writing the balanced <u>net ionic</u> chemical equation of the NaOH reaction.		(I): (R):
A.7			(O):
	Q.A.3 What ion caused the shift in equilibrium in step A.7?		(I):

B. $HIn + OH^- \rightleftharpoons In^- + H_2O$

Color: ______   ______ $\leftarrow$ bromocresol green

Color: ______   ______ $\leftarrow$ phenolphthalein

Step	Stress on System	Direction of Shift	Observations (O) and Interpretations (I)
B.3			(O):
	Q.B.1 Is the bromocresol green indicator in its acid or base form when starting this trial?		(I):
B.5			(O):
	Q.B.2 Explain what is happening in step B.5 in terms of Le Châtelier's Principle.		(I):
B.7			(O):
	Q.B.3 Explain what is happening in step B.7 in terms of Le Châtelier's Principle.		(I):
B.11			(O):
	Q.B.4 Is the phenolphthalein indicator in its acid or base form when starting this trial?		(I):
B.13			(O):
	Q.B.5 Explain what is happening in step B.13 in terms of Le Châtelier's Principle.		(I):
B.15			(O):
	Q.B.6 Explain what is happening in step B.15 in terms of Le Châtelier's Principle.		(I):

C. $Fe^{3+} + SCN^- \rightleftharpoons Fe(SCN)^{2+}$

Ion Color: ______

Step	Stress on System	Direction of Shift	Observations (O), Interpretations (I), and Reactions (R)
C.3			
	Q.C.1 Where does the equilibrium lie in the control test tube?		(I):
C.4			(O):
	Q.C.2 Explain what is happening in step C.4 in terms of Le Châtelier's Principle.		(I):
C.5			(O):
	Q.C.3 Explain what is happening in step C.5 in terms of Le Châtelier's Principle.		(I):
C.6			(O):
	Q.C.4 Explain what the addition of NaOH in this step does to the reaction in terms of Le Châtelier's Principle.		(I):
	Support this by writing the balanced net ionic chemical equation of the NaOH reaction.		(R):
C.7			(O):
	Q.C.5 Explain what the addition of AgNO$_3$ in this step does to the reaction in terms of Le Châtelier's Principle.		(I):
	Support this by writing the balanced net ionic equation of the AgNO$_3$ reaction. The formation of a precipitate is not evidence of a shift in equilibrium because it is not one of the substances in the equilibrium.		(R):

D. $CoCl_4^{2-} + 6\,H_2O \rightleftharpoons 4\,Cl^- + Co(H_2O)_6^{2+} + $ heat

Ion
Color: ______   ________

Step	Stress on System	Direction of Shift	Observations (O) and Interpretations (I)
D.5			(O):
	Q.D.1. Why is the pink cobalt complex called the "aqua" complex?		(I):
	Q.D.2. Explain what is happening in step D.5 in terms of Le Châtelier's Principle.		(I):
	Q.D.3. Explain why the cobalt chloride complex is prepared as a solution in methanol and not water in terms of Le Châtelier's Principle.		(I):
D.6			(O):
	Q.D.4. Explain what is happening in step D.6 in terms of Le Châtelier's Principle.		(I):
D.7			(O):
	Q.D.5. Explain what is happening in step D.7 in terms of Le Châtelier's Principle.		(I):
	Q.D.6. Is this reaction endothermic or exothermic?		(I):

E. $BaCrO_{4\ (s)} \rightleftharpoons Ba^{2+}_{\ (aq)} + CrO_4^{2-}_{\ (aq)}$

Step	Stress on System	Direction of Shift	Observations (O), Interpretations (I), and Reactions (R)
E.4			(O):
E.5			(O):
	Q.E.1. Explain what is happening in step E.5 in terms of Le Châtelier's Principle. Write the balanced net ionic chemical reaction of HCl with a species in the reaction.		(I): (R):

Name___________________________

HOMEWORK EXERCISES

1. State Le Châtelier's Principle.

2. a. List four factors that can shift the position of an equilibrium.

    b. Which one of the above factors can alter the value of the equilibrium constant?

3. Hydrogen is produced from the methane in natural gas as follows:

$$CH_{4\,(g)} + H_2O_{(g)} \rightleftharpoons CO_{(g)} + 3\,H_{2\,(g)} \qquad \Delta H° = 206.2\ kJ$$

The constant volume system is allowed to reach an equilibrium position. A stress is placed on the system. Predict the direction of the shift needed to reestablish equilibrium.

Change or stress on the system at equilibrium	Direction of shift (left or right) to reestablish equilibrium	Change in molar concentrations immediately after the change/stress has been introduced $(\uparrow$ or $\downarrow)$	
		CH_4 and H_2O	CO and H_2
Example: Remove CH_4	$\leftarrow$	$\uparrow$	$\downarrow$
a. Add CH_4			
b. Add H_2			
c. Remove H_2O			
d. Remove CO			
e. Add N_2			
f. Remove H_2			
g. Increase the vessel's pressure			
h. Increase the vessel's volume			
i. Increase the temperature			
j. Decrease the temperature			
k. Add a catalyst			

Continued

4.   The pollutant SO_2 can be removed using the following equilibrium:

$$CaO_{(s)} + SO_{2\,(g)} \rightleftharpoons CaSO_{3\,(s)}$$

$$\Delta H° = -274.9 \text{ kJ}$$

State what would happen to the equilibrium position <u>and the reason</u> of the shift if the following occurred.

a. Some of the $CaSO_3$ were removed from the system?

b. Some CaO were added to the system?

c. The temperature was increased?

d. The pressure is decreased?

e. Write the equilibrium constant expression (K_c) for this reaction.

f. In order to remove as much SO_2 as possible, what would be good operating conditions for this reaction?

    _______________ (high/low) temperature

    _______________ (high/low) pressure

    _______________ (high/low) concentration of ________________

EXPERIMENT 15
ACIDS AND BASES: pH, MOLARITY, AND IONIZATION CONSTANTS

The basic concepts and calculations involved in acid-base titrations is presented. First, a weak acid will be titrated with a strong base, and then a weak base will be titrated with a strong acid. Titration curves of pH vs. mL titrant will be made, and the molarity and K_a or K_b values will be determined. Next, various solutions will be analyzed with a pH meter to determine their molarity, $[H^+]$, and $[OH^-]$ values.

Key Chemical Reactions:

$$CH_3COOH_{(aq)} + NaOH_{(aq)} \rightarrow CH_3COONa_{(aq)} + H_2O_{(l)}$$

$$CH_3COOH_{(aq)} \rightleftharpoons COOH^-_{(aq)} + H^+_{(aq)}$$

$$HCl_{(aq)} + NH_4OH_{(aq)} \rightarrow NH_4Cl_{(aq)} + H_2O_{(l)}$$

$$NH_{3\,(aq)} + H_2O_{(l)} \rightleftharpoons NH_4^+{}_{(aq)} + OH^-_{(aq)}$$

Key Mathematical Equations:

$$pH = -\log[H^+]$$

$$K_w = [H^+][OH^-] = 1 \times 10^{-14}$$

$$K_a = \frac{[A^-][H^+]}{[HA]} \qquad K_b = \frac{[BH^+][OH^-]}{[B]}$$

$$\% \text{ ionization} = \frac{[H^+]}{[HA]} \times 100$$

$$0.0500 \text{ L HA} \times \frac{0.125 \text{ mole HA}}{L} \times \frac{1 \text{ mole NaOH}}{1 \text{ mole HA}} \times \frac{1000 \text{ mL}}{0.415 \text{ mole NaOH}} = 15.1 \text{ mL NaOH}$$

Discussion
Acids and Bases
An acid is a substance that ionizes in water to give H^+ ions, and a base is a substance that ionizes in water to give OH^- ions. In more general terms, a Brønsted acid is a proton donor and a Brønsted base is a proton acceptor. For example, formic acid, HCOOH, is considered an acid because it donates a proton to water in the following equation:

$$HCOOH_{(aq)} + H_2O_{(l)} \rightleftharpoons COOH^-_{(aq)} + H_3O^+_{(aq)}$$

Here, the $COOH^-$ on the product side is termed the conjugate base of HCOOH, and H_3O^+ is termed the conjugate acid of H_2O.

Similarly, ammonia, NH_3, is considered a base because it accepts a proton:

$$NH_{3\ (aq)}\ +\ H_2O_{\ (l)}\ \rightleftharpoons\ NH_4^+{}_{\ (aq)}\ +\ OH^-{}_{\ (aq)}$$

Here, the NH_4^+ on the product side is termed the conjugate acid of NH_3, and OH^- is termed the conjugate base of H_2O.

The conjugate acids and bases (H_3O^+ and OH^-) are of importance in these equations.

pH
A measure of the acidity of a solution is defined as the concentration of hydrogen ions (H_3O^+, or more commonly H^+) in solution. Since this value is a very small number, the "p" of the value is taken, where $p = -\log_{10}$. Therefore, the pH of a solution is defined as follows:

$$pH = -\log[H^+]$$

Similarly, the pOH of a solution is defined as $-\log[OH^-]$. The corollary equations of these are

$$[H^+] = 10^{-pH} \qquad \text{and} \qquad [OH^-] = 10^{-pOH}$$

Acid Strength
The following equation can be written for the ionization of water:

$$H_2O_{\ (l)}\ \rightleftharpoons\ H^+{}_{\ (aq)}\ +\ OH^-{}_{\ (aq)}$$

By definition, the ionization constant expression for water (K_w) at 25 °C is written

$$K_w = [H^+][OH^-] = 1 \times 10^{-14}$$

Therefore,
   when $[H^+] = 1 \times 10^{-7}$, the pH $= 7$ and the solution is neutral
   when $[H^+] > 1 \times 10^{-7}$, the pH < 7 and the solution is acidic
   when $[H^+] < 1 \times 10^{-7}$, the pH > 7 and the solution is basic.

Also, by definition,
$$pH + pOH = 14$$

It is important to understand that since this is a logarithmic scale, lowering the pH by one unit means the solution becomes ten times more acidic.

The strength of an acid can be determined a number of ways: by its pH, molarity, acid ionization constant (K_a), and by % ionization. Each of these will be discussed, as well as the situations in which they can be used for comparison.

pH
The strength of acid solutions can be compared based on pH if they meet one of the following criteria: they must be either solutions of the same acid of different pH or different acids having the same concentration. When comparing solutions containing the same acid, the solution with the lower pH is stronger. For example, an acetic acid

solution of pH 2.65 is stronger than an acetic acid solution having a pH of 2.83. When comparing the same concentrations of different acids, the lower the pH, the stronger the acid. For example, a 0.20 M formic acid solution having a pH of 2.23 is stronger than a 0.20 M acetic acid solution having a pH of 2.72.

Molarity
Solutions of the same acid can also be evaluated based on molarity: the higher the molarity, the stronger the acid. If two solutions of acetic acid are compared, a 1.25 M solution of acetic acid ($[H^+]$ = 1.25 M; pH = 0.10) is stronger than a 0.125 M solution ($[H^+]$ = 0.125 M; pH = 0.90) because it has a greater molarity. Comparing the pH values of the same acid will also give the same result, as noted in the pH example.

K_a Values
Acids can also be compared based on their degree of ionization. Strong acids and strong bases completely dissociate in water (one-way arrow, $\rightarrow$). There are six strong acids (HCl, HI, HBr, H_2SO_4, HNO_3, and $HClO_4$), and the strong bases are the Group I and II hydroxides. For example, HCl completely dissociates according to the equation

$$HCl_{(aq)} \rightarrow H^+_{(aq)} + Cl^-_{(aq)}$$

Strong acids and strong bases do not have K_a or K_b values.

Weak acids and bases, however, only partially dissociate in water. At equilibrium (note the $\rightleftharpoons$ arrow in the following equation), there will be mostly the weak acid present, but there will be a small amount of H^+ (H_3O^+) in solution. For a generic weak acid, HA,

$$HA_{(aq)} + H_2O_{(l)} \rightleftharpoons A^-_{(aq)} + H_3O^+_{(aq)}$$

the solution will contain mostly HA, but there will be a small amount of H^+ present which can be measured by pH.

The ionization constant expression for the above dissociation can be written using K_a, the acid ionization constant, as follows:

$$K_a = \frac{[A^-][H^+]}{[HA]}$$

Therefore, when comparing acids, as a K_a increases, the strength of the acid increases because $[H^+]$ increases. For example, HF, which has a K_a of 7.1 x 10^{-4}, is a stronger acid than acetic acid, which has a K_a of 1.8 x 10^{-5}. A K_a value is unique and constant for a weak acid, regardless of its concentration.

% Ionization
Finally, the strengths of different acids of the same concentration can be evaluated based on percent ionization. The percent ionization is given by the equation

$$\% \text{ ionization} = \frac{[H^+]}{[HA]} \times 100$$

The acid with the lower pH (higher [H⁺]) has the higher percent ionization, and is therefore stronger. The strengths of acids can only be compared in the situation of different acids of the same concentration; it cannot be used to determine the strength of the same acid of different concentrations. In this latter case, the acid with the lower concentration (more dilute acid), has a higher percent ionization.

A summary of comparing acid strengths based on different criteria is given in Table 15.1.

Table 15.1. Acid Strength Criteria

Criteria	Situation	What Gives a Stronger Acid
pH	Same acid of different pH **OR** Same concentration of different acids	Lower pH = stronger acid
M	Same acid of different concentration	Higher M = stronger acid
K_a	Different acids of varying concentrations	Higher K_a = stronger acid
% Ionization	Same concentration of different acids	Higher % ionization = stronger acid
	Note: The strength of *same* acids of different concentrations cannot be compared by % ionization	

The Acid-Base Titration Curve
A typical weak acid-strong base titration curve is shown in Figure 15.1.

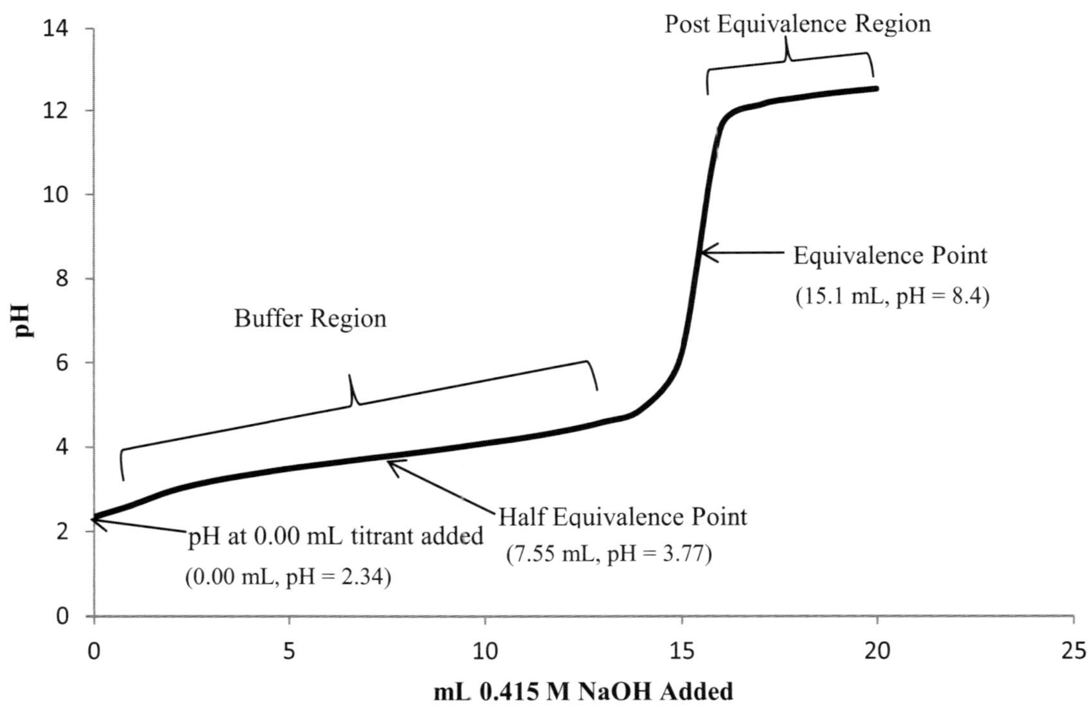

Figure 15.1
Weak Acid-Strong Base Titration Curve

There are three important points on this curve and two important regions.

- The point <u>pH at 0.00 mL titrant added</u> is the point at which no base has been added, where the curve crosses the y axis. This is also known as the pH at initial concentration.
- The <u>equivalence point</u> is where the moles of acid present are stoichiometrically equal to the moles of base added. At this point, there is no acid or base in solution, but only salt and water.
- The <u>half equivalence point</u> is where exactly half of the moles of acid have been reacted with the base, and volume-wise it is exactly half of the volume needed to reach the equivalence point.
- The <u>buffer region</u> is a flat region where the weak acid and its salt are present, and the pH is "buffered" – there is very little change in pH when a large amount of base is added.
- The <u>post equivalence region</u> is the region after the equivalence point to the endpoint.

Diprotic acids (two H^+) have two buffer regions, two equivalence points, and two half equivalence points. Triprotic acids (three H^+) have three buffer regions, three equivalence points, and three half equivalence points. However, some equivalence points, usually the third, are often difficult to detect.

Weak bases have exactly the opposite shaped curve, but the same important points and regions.

Calculation of the Three Noted Points on a Titration Curve
This is the first of two labs that detail how to calculate certain values in an acid base titration curve.

The calculations will be described in the following example:
50.0 mL of a 0.125 M solution of HCOOH (K_a = 1.7 x 10^{-4}) are titrated with 0.415 M NaOH.

<u>pH at Initial Concentration, Molarity, and K_a</u>
This is the first point on the curve where 0.00 mL of NaOH have been added. In this calculation, there are three variables: the K_a of the acid, the pH, and the acid molarity. The chemical equation of the weak acid dissociation is as follows.

$$HCOOH_{(aq)} \rightleftharpoons COOH^-_{(aq)} + H^+_{(aq)}$$

The first step is to set up an ICE table specific to the reaction. An ICE table displays the initial, change and equilibrium concentrations at a point in the reaction. For the initial concentrations, the only component in solution is the acid, and in this example its molarity is known. For the change concentrations, it is assumed that the acid dissociates to a concentration "x", and at the same time, the $COOH^-$ and the H^+ increase in concentration by the same amount. If there is a stoichiometric coefficient in front of a component, this must be included as a multiplier with the "x". (If there is a 2 A^-, then the

"C" will be 2x) The equilibrium concentration for each component is the addition of the initial and change values.

$$HCOOH_{(aq)} \rightleftharpoons COOH^-_{(aq)} + H^+_{(aq)}$$

I	0.125	0	0
C	−x	x	x
E	0.125 − x	x	x

The ionization constant expression for the above dissociation can be written using K_a, the acid ionization constant, as follows:

$$K_a = \frac{[COOH^-][H^+]}{[HCOOH]}$$

Substituting in the equilibrium values,

$$1.7 \times 10^{-4} = \frac{[x][x]}{[0.125 - x]}$$

Solving for x (the x in the denominator can be taken to 0, avoiding the quadratic, because x <<< 0.125) gives x = 0.0046. It should be realized that this value is equal to $[H^+]$. Taking the −log of this value gives the pH, 2.34.

In a similar manner, the K_a can be calculated if the initial acid concentration and pH are known, or the initial concentration can be calculated if the pH and the K_a are known.

For weak bases ($B \rightleftharpoons BH^+ + OH^-$), the ionization constant expression is

$$K_b = \frac{[BH^+][OH^-]}{[B]}$$

K_a and K_b can be calculated from one another by the equation $K_w = K_a K_b$.

<u>Equivalence Point</u>
The next point is the equivalence point. The mL to the equivalence point is calculated from the molarity and volume of the acid and the molarity of the base. This is shown in the following equation:

$$0.0500 \text{ L HCOOH} \times \frac{0.125 \text{ mole HCOOH}}{\text{L}} \times \frac{1 \text{ mole NaOH}}{1 \text{ mole HCOOH}} \times \frac{1000 \text{ mL}}{0.415 \text{ mole NaOH}} = 15.1 \text{ mL NaOH}$$

For this lab, the pH at this volume can be visually estimated from the graph. The point is located halfway between the two points of inflection in the sharp incline of the curve. Referring to Figure 15.1, the pH is 8.4 at 15.1 mL of NaOH added. The method to calculate the exact pH at the equivalence point will be demonstrated in the following lab.

Half Equivalence Point and K_a

The last point that will be addressed is the half equivalence point, which is the point at which half of the acid has reacted. The volume to the half equivalence point is exactly half of the volume to the equivalence point, $\frac{1}{2}(15.10 \text{ mL}) = 7.55 \text{ mL}$. The pH at the half equivalence point can also be determined by knowing that at this point, $pH = pK_a$. In this example, since the $K_a = 1.7 \times 10^{-4}$, the p of this value gives 3.77, the pH at the half equivalence point. Similarly, $pOH = pK_b$ at the $\frac{1}{2}$ equivalence point.

Chemicals

    Acetic acid, CH_3COOH
    Ammonium hydroxide, NH_4OH
    Bromocresol green
    Hydrochloric acid, HCl
    Phenolphthalein
    Sodium hydroxide, NaOH
    Various household chemicals

Waste

    All waste can be disposed down the laboratory sink.

Procedure

Titration of Acetic Acid with NaOH

$$CH_3COOH + NaOH \rightarrow CH_3COONa + H_2O$$

1. Calibrate the pH meter according to the instructions in the Appendix. Unscrew the pH probe from the storage container (if applicable). Leave this top on during the calibration and all titrations.

2. Normalize a buret with the titrating NaOH solution. Drain the normalizing liquid to the drain and fill the buret.

3. Dispense 50.0 mL of the acetic acid solution to be titrated to a clean and dry 250 mL beaker. Add 2-3 drops of phenolphthalein.

4. Gently shake off any liquid on the pH probe, rinse the tip and body with DI water, and then blot the whole electrode with a paper towel to completely dry it. Insert the probe into the acetic acid. Allow the initial pH to stabilize about five minutes before taking a reading. Leave the probe in the beaker during the titration. It can be used to gently stir the solution during the titration.

5. Titrate the acetic acid with the NaOH, adding the NaOH in 1 mL increments. (Although the exact volume of 1 mL each time is not required, doing so will make this titration proceed more quickly, and is quite adequate for the basic calculations of this lab.) Monitor and record the pH after each increment of NaOH has been added. Titrate until the final pH is above 12 for at least three pH readings.

6. Graph the data obtained and complete the calculations.

EXPERIMENT 15   ACIDS AND BASES: pH, MOLARITY, K_a AND K_b

Titration of Ammonium Hydroxide with HCl

$$HCl + NH_4OH \rightarrow NH_4Cl + H_2O$$

1. Normalize a buret with the titrating HCl solution. Drain the normalizing liquid to the drain and fill the buret.

2. Dispense 50.0 mL of the ammonium hydroxide solution to be titrated to a clean and dry 250 mL beaker. Add 5-6 drops of bromocresol green.

3. Gently shake off any liquid on the pH probe, rinse the tip and body with DI water, and then blot the whole electrode with a paper towel to completely dry it. Insert the probe into the ammonium hydroxide. Allow the initial pH to stabilize about five minutes before taking a reading. Leave the probe in the beaker during the titration. It can be used to gently stir the solution during the titration.

4. Titrate the ammonium hydroxide with the HCl, adding the HCl in 1 mL increments. Monitor and record the pH after each increment of HCl has been added. Titrate until the final pH is below 2 for three pH readings.

5. Graph the data obtained and complete the calculations.

pH, Molarity and Strength of Common Household Solutions
1. Obtain approximately 30 mL of each solution in a 50 mL beaker.

2. Measure the pH of each solution and complete the data table. Be sure to normalize the electrode between each measurement by rinsing the probe with a small amount of the new solution, and then gently shaking it off.

EXPERIMENT 15   ACIDS AND BASES: pH, MOLARITY, K_a AND K_b

LABORATORY REPORT SHEET: Acetic Acid ($HC_2H_3O_2$) and NaOH

Name ________________________________ Date ____________ M NaOH __________

Buret Reading, mL	Total mL NaOH added	pH		Buret Reading, mL	Total mL NaOH added	pH		Buret Reading, mL	Total mL NaOH added	pH
*										

*   This initial reading should be subtracted from subsequent buret readings (column 1) to obtain the total mL of NaOH added (column 2).

1. Calculate the molarity of the acetic acid solution using an ICE table and the ionization constant expression. K_a, acetic acid = 1.8×10^{-5}.

2. Calculate the mL of NaOH added to reach the equivalence point based on the volume and molarity of the acid.

3. Graph the data. By inspection of the graph, what are the pH and mL NaOH added at the equivalence point?

    pH __________   mL NaOH added ________________

4. Using the data from (3), calculate the volume of NaOH to the half equivalence point.

5. Calculate K_a based on the half equivalence point data of the graph.

6. Clearly mark on the graph the mL and pH values at 0.00 mL NaOH added, ½ EP, and EP from questions 1, 3, 4, and 5 above. Write in the form (___ mL, pH = ___).

LABORATORY REPORT SHEET: _______ M NH_3 and ______ M HCl

Buret Reading, mL	Total mL HCl added	pH		Buret Reading, mL	Total mL HCl added	pH		Buret Reading, mL	Total mL HCl added	pH
*										

*   This initial reading should be subtracted from subsequent buret readings (column 1) to obtain the total mL of HCl added (column 2).

1.   Calculate the K_b of ammonia using an ICE table and the ionization constant expression.

2.   Calculate the mL of HCl needed to reach the equivalence point based on the volume and molarity of the base.

3.   Graph the data. By inspection of the graph, what are the pH and mL HCl added at the equivalence point?

        pH _________   mL HCl added _______________

4.   Using the data from (3), calculate the volume of HCl to the half equivalence point.

5.   Calculate K_b based on the half equivalence point data of the graph.

6.   Clearly mark on the graph the mL and pH values at 0.00 mL HCl added, ½ EP, and EP from the starting pH (table) and questions 3, 4, and 5 above. Write in the form (___ mL, pH = ___).

LABORATORY REPORT SHEET: pH and Data of Solutions   Name_______________________________

Solution ID	pH at initial concentration	$[H^+]$	M	$[OH^-]$	K_a	K_b
For acetic acid, record the pH at initial concentration from the data, and the M and literature value of K_a from Question 1.						
Acetic acid (first titration)						
Store "Vinegar"						
For NH$_4$OH, record the pH at initial concentration from the data, the M as given, and the K_b value from the ½ equivalence point of the trial (Question 5).						
NH$_4$OH (second titration)						
Window or glass cleaner (with Ammonia)						
Store "Ammonia"						

Remember: K_a and K_b are <u>specific and constant</u> for an acid or base, independent of the concentration of the acid or base.

1. For the two solutions that contain acetic acid, rank the acids based on strength. Explain on what basis the determination was made.

   __________________ $<$ __________________

   Explain:

2. For the two solutions that contain acetic acid, calculate the % ionization of each. Which has the larger % ionization, and why? (The seemingly puzzling answer will make clear why % ionization can only be used to compare solutions of the same concentration.)

3. For the three solutions that contain ammonia, rank them based on basic strength.

   __________________ $<$ __________________ $<$ __________________

   Explain on what basis the determination was made:

4. In what situations can pH be used to determine the acid strength of a solution?

5. In what situation can M be used to determine the acid strength of a solution?

6. In what situation can K_a be used to determine the acid strength of a solution?

7. In what situation can % ionization be used to determine the acid strength of a solution?

<u>HOMEWORK EXERCISES</u>   Name ___________________________

For Questions 1-4, write an equation for the dissociation of the weak acid, set up an ICE table and ionization constant expression, and then calculate the parameters on the right. For these questions, no titrant has been added.

1. 0.275 M HCOOH; $K_a = 1.7 \times 10^{-4}$

Equation: $HCOOH \rightleftharpoons COOH^- + H^+$

ICE Table: $[H^+]$: ______________

                                                               pH: ______________

                                                             pOH: ______________

                                                              $[OH^-]$: ______________

                                                               K_b: ______________

2. 0.275 M C_6H_5COOH; pH = 2.37

Equation:

ICE Table: $[H^+]$: ______________

                                                              K_a: ______________

                                                                pOH: ______________

                                                              $[OH^-]$: ______________

                                                                K_b: ______________

3. CH_3COOH; pH = 2.62; $K_a = 1.8 \times 10^{-5}$

Equation:

ICE Table: $[H^+]$: ______________

                                                               M: ______________

                                                                pOH: ______________

                                                                $[OH^-]$: ______________

                                                                K_b: ______________

4. 0.275 M CH_3COOH; $K_a = 1.8 \times 10^{-5}$

Equation:

ICE Table: [H^+]: ___________

                                                                              pH: ___________

                                                                              pOH: ___________

                                                                              [OH^-]: ___________

                                                                              K_b: ___________

5. Calculate the percent ionization of the acids in questions 1-4:

0.275 M HCOOH	0.275 M C_6H_5COOH	_______ M CH_3COOH	0.275 M CH_3COOH

For Questions 6-9, refer to Table 15.1 for the "Situation" when each is applicable.

6. Three acids in 1-4 can be ranked based on pH. Rank them in order of increasing acidity (list the molarity and identity of the acid) and state the specific situation when this can be used.

    _______________ < _______________ < _______________

    Situation:

7. Three acids in 1-4 can be ranked based on K_a value. Rank them in order of increasing acidity (include the molarity and identity of the acid) and state the specific situation when this can be used.

    _______________ < _______________ < _______________

    Situation:

8. Three different acids in 1-4 can be ranked based on % ionization. Rank them in order of increasing acidity (include the molarity and identity of the acid) and state the specific situation when this can be used.

    _______________ < _______________ < _______________

    Situation:

9. The two CH_3COOH solutions can be compared. Rank them in order of increasing acidity (include the molarity and identity of the acid) and list two specific situations when these can be used.

    _______________ < _______________ Situations:

For Question 10, write an equation for the dissociation of the weak base, set up an ICE table and ionization constant expression, and then calculate the parameters on the right. For this question, no titrant has been added.

10.   0.600 M NH_3; $K_b = 1.8 \times 10^{-5}$

Equation:

ICE Table: $[H^+]$: _______________

                                                                                        pH: _______________

                                                                                        pOH: _______________

                                                                                        $[OH^-]$: _______________

                                                                                        K_a: _______________

11.   50.0 mL of 0.325 M CH_3COOH ($K_a = 1.8 \times 10^{-5}$) are titrated with 0.482 M NaOH.
    a. Write the balanced chemical equation for this reaction.

    b. Calculate the mL of NaOH needed to reach the equivalence point.

    c. Calculate the mL of NaOH needed to reach the half equivalence point.

    d. Calculate the pH of the acid at the half equivalence point.

12.   50.0 mL of 0.325 M CH_3COOH ($K_a = 1.80 \times 10^{-5}$) are titrated with 0.482 M $Ba(OH)_2$.
    a. Write the balanced chemical equation for this reaction.

    b. Calculate the mL of $Ba(OH)_2$ needed to reach the equivalence point.

    c. Calculate the mL of $Ba(OH)_2$ needed to reach the half equivalence point.

    d. Calculate the pH of the acid at the half equivalence point.

EXPERIMENT 15   ACIDS AND BASES: pH, MOLARITY, K_a AND K_b

EXPERIMENT 16
ACIDS AND BASES: COMPLETE CURVE ANALYSIS INCLUDING BUFFERS, EQUIVALENCE POINTS, AND POST EQUIVALENCE REGIONS

Three titrations will be performed: a weak acid with a strong base, a polyprotic weak acid with a strong base, and a strong acid with a strong base. The changes in pH during the titrations will be monitored and a curve for each titration will be plotted. The exact pH at various points along the titration curve (the pH at initial acid concentration where 0.00 mL of titrant have been added, a point in the buffer region, the equivalence point, and a point in the post equivalence region) will be calculated. The ionization constants for the weak acids will also be calculated.

Key Chemical Reactions:

$$KHP \; + \; NaOH \; \rightarrow \; NaKP \; + \; H_2O$$
$$HC_2H_3O_2 \; + \; NaOH \; \rightarrow \; NaC_2H_3O_2 \; + \; H_2O$$
$$H_3PO_4 \; + \; 3\,NaOH \; \rightarrow \; Na_3PO_4 \; + \; 3\,H_2O$$
$$HCl \; + \; NaOH \; \rightarrow \; NaCl \; + \; H_2O$$

$HA \; \rightleftharpoons \; H^+ \; + \; A^-$	weak acid dissociation equation
$HA \; + \; NaOH \; \rightarrow \; NaA \; + \; H_2O$	weak acid + base reaction equation
$NaA \; \rightarrow \; Na^+ \; + \; A^-$	salt dissociation equation
$A^- \; + \; H_2O \; \rightleftharpoons \; HA \; + \; OH^-$	salt hydrolysis equation

$B \; + \; H_2O \; \rightleftharpoons \; BH^+ \; + \; OH^-$	weak base dissociation equation
$B \; + \; HCl \; \rightarrow \; BHCl$	acid + weak base reaction equation
$BHCl \; \rightarrow \; BH^+ \; + \; Cl^-$	salt dissociation equation
$BH^+ \; + \; H_2O \; \rightleftharpoons \; B \; + \; H_3O^+$	salt hydrolysis equation

Key Mathematical Equations:

$$pH = -\log[H^+] \qquad\qquad K_w = [H^+][OH^-] = 1 \times 10^{-14}$$

$$K_a = \frac{[A^-][H^+]}{[HA]} \qquad\qquad K_b = \frac{[BH^+][OH^-]}{[B]}$$

$$0.0500\ \text{L HA} \times \frac{0.125\ \text{mole HA}}{\text{L}} \times \frac{1\ \text{mole NaOH}}{1\ \text{mole HA}} \times \frac{1000\ \text{mL}}{0.415\ \text{mole NaOH}} = 15.1\ \text{mL NaOH}$$

ICE Tables

	$HA_{(aq)}$	$\rightleftharpoons$	$A^-_{(aq)}$	+	$H^+_{(aq)}$
I	0.0350		0.0692		0
C	$-x$		x		x
E	$0.0350 - x$		$0.0692 + x$		x

<u>Discussion and More Calculations in a Monoprotic Acid Titration Curve</u>
The example from the previous experiment will be continued. In this example, shown again as Figure 16.1, 50.0 mL of 0.125 M HCOOH are titrated with 0.412 M NaOH. The K_a for HCOOH is 1.7×10^{-4}.

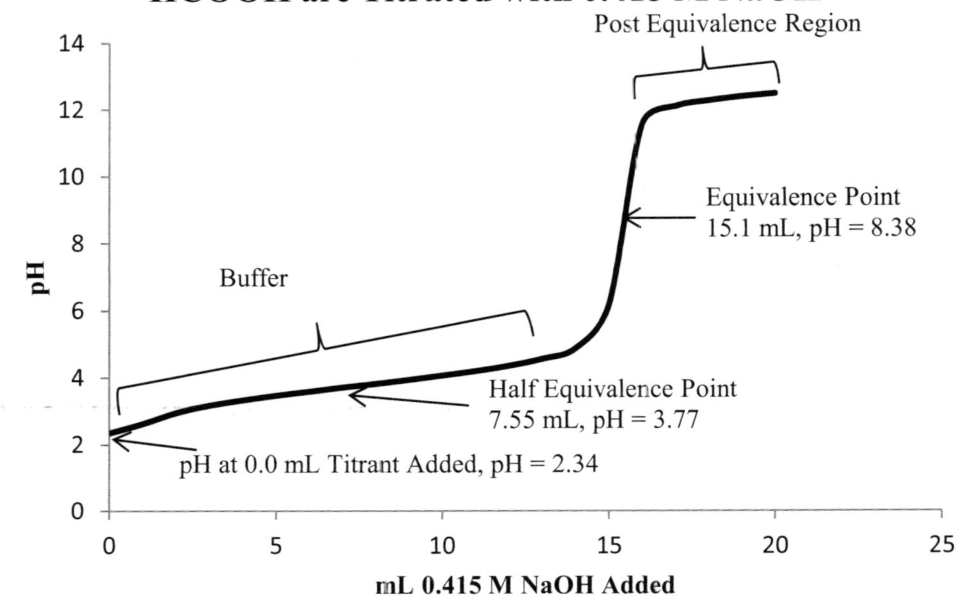

Figure 16.1
Titration of 50.0 mL of 0.125 M HCOOH with 0.415 M NaOH

The titration reaction is as follows:

$$HCOOH_{(aq)} + NaOH_{(aq)} \rightarrow NaCOOH_{(aq)} + H_2O_{(l)}$$

This reaction is assumed to go to completion where each mole of NaOH added reacts with a mole of HCOOH and forms a mole of $COOH^-$, in the form of NaCOOH.

There will be five points calculated on this curve:
- pH at 0.0 mL of titrant added
- pH at a point in the buffer region
- Exact pH at the equivalence point
- pH at a point in the post equivalence region
- pH at the half equivalence point

These different points each have a specific method to calculate the associated pH values.

pH at 0.00 mL Titrant Added
This was calculated in the previous experiment, but will be summarized here.

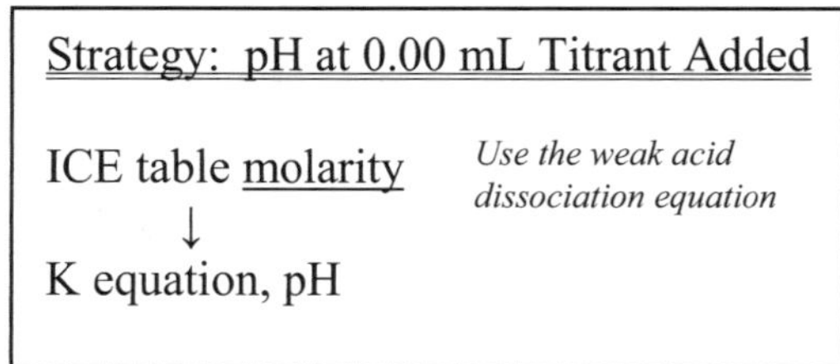

1. The weak acid dissociation equation is written, and an ICE table using the initial molarity of the acid is set up.

$$HCOOH_{(aq)} \rightleftharpoons COOH^-_{(aq)} + H^+_{(aq)}$$

	HCOOH	COOH⁻	H⁺
I	0.125	0	0
C	$-x$	x	x
E	$0.125 - x$	x	x

2. The ionization constant expression for the above dissociation can be written using K_a, the acid ionization constant, as follows:

$$K_a = \frac{[COOH^-][H^+]}{[HCOOH]} \qquad 1.7 \times 10^{-4} = \frac{[x][x]}{[0.125 - x]}$$

3. Solve for x and calculate the pH.
    The x in the denominator can be taken to 0, avoiding the quadratic, because x $<<<$ 0.125, giving x = 0.0046 M.

$$x = [H^+] = 4.6 \times 10^{-3}\,M$$

$$pH = -\log[H^+] = 2.34$$

pH in the Buffer Region
In the buffer region, the value of the pH changes very little (it is "buffered") due to the presence of a weak acid and its salt, or a weak base and its salt. To calculate the pH, first an ICE table based on <u>moles</u> is set up to determine what reacted and what is left in solution. Then, a new molarity of each component still present in solution will be calculated based on the new total volume. A molar ICE table will be set up using the weak acid dissociation equation to calculate $[H^+]$, and a pH can be determined from this value.

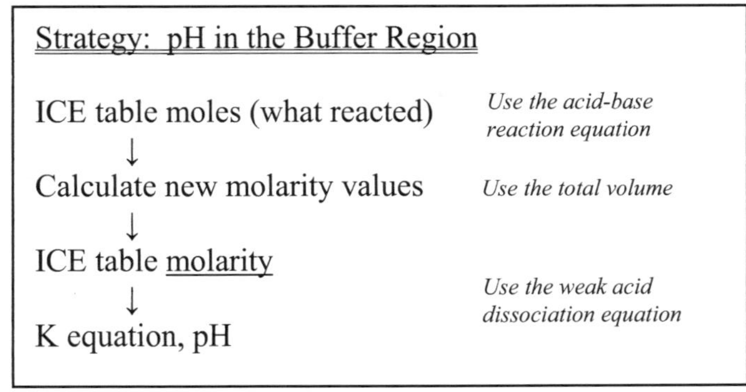

Calculate the pH when 10.00 mL of NaOH are added to the HCOOH solution.

1. Determine what has reacted and what is still in solution after the reaction using an ICE moles table and the acid-base reaction equation.

   It can clearly be seen that NaOH is the limiting reagent: no NaOH will be left in solution after the reaction.

$$HCOOH_{(aq)} \quad + \quad NaOH_{(aq)} \quad \longrightarrow \quad NaCOOH_{(aq)} + \quad H_2O_{(l)}$$

	HCOOH	NaOH	NaCOOH
I	(0.0500 L)(0.125 M)	(0.01000 L)(0.415 M)	0
C	−0.00415 moles	−0.00415 moles	+0.00415 moles
E	0.00210 moles	0 moles	0.00415 moles

2. Calculate a new molarity for the HCOOH and NaCOOH.

   The new total volume is (50.0 mL + 10.00 mL) = 60.0 mL, or 0.0600 L.

$$HCOOH_{(aq)} \quad + \quad NaOH_{(aq)} \quad \longrightarrow \quad NaCOOH_{(aq)} + \quad H_2O_{(l)}$$

$$\frac{0.00210 \text{ moles}}{0.0600 \text{ L}} \qquad\qquad\qquad \frac{0.00415 \text{ moles}}{0.0600 \text{ L}}$$

$$= 0.0350 \text{ M} \qquad\qquad\qquad = 0.0692 \text{ M}$$

3. Set up an ICE table based on the new molarity values using the acid dissociation equation.

$$HCOOH_{(aq)} \quad \rightleftharpoons \quad COOH^-_{(aq)} + \quad H^+_{(aq)}$$

	HCOOH	COOH⁻	H⁺
I	0.0350	0.0692	0
C	−x	x	x
E	0.0350 − x	0.0692 + x	x

4. Set up the ionization constant expression for the above equilibrium values using K_a, the acid ionization constant, since H^+ is in the products.

$$K_a = \frac{[COOH^-][H^+]}{[HCOOH]} \quad \text{or} \quad 1.7 \times 10^{-4} = \frac{[0.0692 + x][x]}{[0.0350 - x]}$$

5. Solve for x and calculate the pH.
   The x in the denominator can be taken to 0 as well as the "+x" in the [0.0692 + x], avoiding the quadratic, because x is very small compared to 0.0350 and 0.0692. This gives x = 8.60 × 10⁻⁵ M.

$$x = [H^+] = 8.60 \times 10^{-5} \text{ M}$$

$$pH = -\log[H^+] = 4.06$$

pH at the Equivalence Point

At the equivalence point of an acid-base titration, no acid or base is present, and the pH is determined by the identity of the salt in the solution. To calculate the pH, first the volume of NaOH needed to reach the equivalence point is calculated. Next, an ICE table based on <u>moles</u> is set up to determine what reacted and what is left in solution (no acid or base should be present; only salt). Then, a new molarity of the salt present in solution will be calculated based on the new total volume. An ICE table will be set up using the salt hydrolysis equation to calculate $[OH^-]$ (in this case), and a pH can be determined from this value. The following calculations outline this method to determine the pH at the equivalence point in this example.

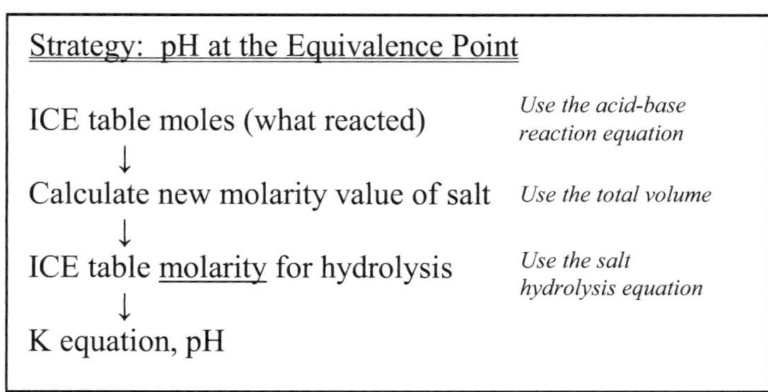

1. Calculate the volume of NaOH needed to reach the equivalence point.

$$0.0500 \text{ L HCOOH} \times \frac{0.125 \text{ mole HCOOH}}{\text{L}} \times \frac{1 \text{ mole NaOH}}{1 \text{ mole HCOOH}} \times \frac{1 \text{ L}}{0.415 \text{ mole NaOH}} = 0.0151 \text{ L}$$

2. Determine what has reacted and what is still in solution after the reaction using an ICE moles table and the acid-base reaction equation. It should be noted that stoichiometrically HCOOH and NaOH are equal, and will not appear in the equilibrium.

$$HCOOH_{(aq)} \quad + \quad NaOH_{(aq)} \quad \rightarrow \quad NaCOOH_{(aq)} \quad + \quad H_2O_{(l)}$$

	HCOOH	NaOH	NaCOOH
I	(0.0500 L)(0.125 M)	(0.0151 L)(0.415 M)	0
C	−0.00625 moles	−0.00625 moles	+0.00625 moles
E	0 moles	0 moles	0.00625 moles

3. Calculate a new molarity for the NaCOOH. The new total volume is (50.0 mL + 15.1 mL) = 65.1 mL, or 0.0651 L.

$$HCOOH_{(aq)} \quad + \quad NaOH_{(aq)} \quad \rightarrow \quad NaCOOH_{(aq)} \quad + \quad H_2O_{(l)}$$

$$\frac{0.00625 \text{ moles}}{0.0651 \text{ L}}$$

$$= 0.0960 \text{ M}$$

Note: $[COOH^-] = [NaCOOH] = 0.0960 \text{ M}$

4. Set up an ICE table based on the new molarity of the salt and the salt hydrolysis. Note that salt hydrolysis means reacting the salt with water. In this equation, the weak acid will be "regenerated" on the product side.

$$COOH^-_{(aq)} + H_2O_{(l)} \rightleftharpoons HCOOH_{(aq)} + OH^-_{(aq)}$$

I	0.0960	0	0
C	−x	x	x
E	0.0960 − x	x	x

5. Set up the ionization constant expression for the above equilibrium values using K_b (because OH^- is generated on the product side).

$$K_b = \frac{[HCOOH][OH^-]}{[COOH^-]} \quad \text{or} \quad 5.9 \times 10^{-11} = \frac{[x][x]}{[0.0960 - x]}$$

6. Solve for x and calculate the pH.
    The x in the denominator can be taken to 0, because x is very small compared to 0.0960. This gives x = 2.38×10^{-6} M.

$$x = [OH^-] = 2.38 \times 10^{-6}\,M$$

$$pOH = -\log[OH^-] = 5.62$$

$$pH = 8.38$$

pH in the Post Equivalence Region

In the post equivalence region, the value of the pH changes very little because the strong base is present. The pH is dependent upon the volume and molarity of the base. To calculate the pH, first the excess volume of base past the equivalence point is calculated. This is converted to moles using the molarity of the base. A new molarity based on the total volume is calculated and $[OH^-]$ is determined, and pH can be calculated from this.

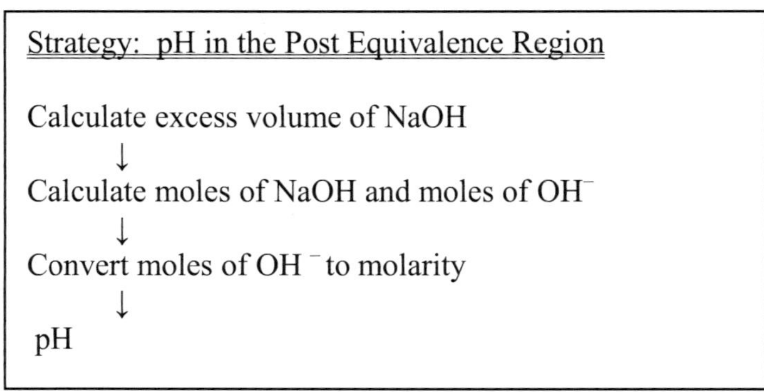

In this example, calculate the pH when 19.0 mL of NaOH have been added.

1. Calculate the excess volume of NaOH present.

    19.0 mL added – 15.1 mL to the equivalence point = 3.9 mL excess

2. Convert this volume to moles of NaOH using the NaOH molarity.

$$3.9 \text{ mL} \times \frac{0.415 \text{ mole NaOH}}{1000 \text{ L}} = 1.61 \times 10^{-3} \text{ moles NaOH} = \text{moles OH}^-$$

3. Calculate a new molarity of OH^-.
    The total volume now is 69.0 mL (50.0 + 19.0).

$$[OH^-] = \frac{1.61 \times 10^{-3} \text{ moles OH}^-}{0.0690 \text{ L}} = 0.0233 \text{ M}$$

4. Calculate a pH.

$$x = [OH^-] = 2.33 \times 10^{-2} \text{ M}$$

$$pOH = -\log[OH^-] = 1.63, \text{ and pH} = 12.37$$

pH at the Half Equivalence Point
The half equivalence point is the point at which half of the acid has reacted. The volume to the half equivalence point is exactly half of the volume to the equivalence point, ½(15.1 mL) = 7.55 mL.

At the half equivalence point, $pH = pK_a$.

$$K_a = 1.7 \times 10^{-4}$$

$$pH_{½ \, EP} = pKa = -\log 1.7 \times 10^{-4} = 3.77$$

Similarly, $pOH = pK_b$ at the ½ equivalence point.

Polyprotic Acids

In the case of polyprotic acids, the titration curve shows an equivalence point for the neutralization of each available H^+. Volumetrically, it requires the same amount of base to remove each H^+.

The points of a polyprotic acid titration will be discussed using the following example: 50.0 mL of sulfurous acid, H_2SO_3, a diprotic acid, are titrated with 0.400 M NaOH as shown in Figure 16.2.

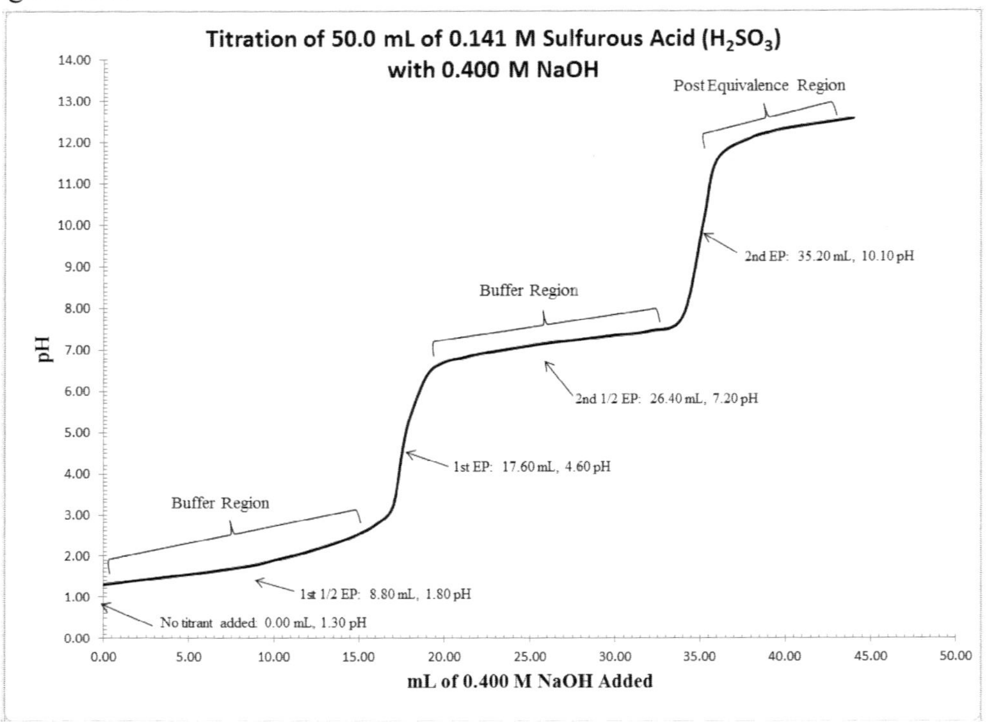

Figure 16.2
Titration of 50.0 mL Sulfurous Acid (H_2SO_3) with 0.400 M NaOH

The mL and pH values at 0.00 mL titrant added, the half equivalence points, and the equivalence points can all be estimated from the graph. Exactly half of the volume to the equivalence point is the volume needed to reach the half equivalence point. The exact same volume which was needed to reach the first equivalence point (removing the first H^+) will be the additional volume needed to reach the second equivalence point (removing the second H^+). Exactly half way volumetrically between the two equivalence points is the second half equivalence point.

Initial			Equivalence Points			½ Equivalence Points	
pH	[H^+]		mL	pH		mL	pH
		First	17.60 mL	4.60	First	8.80 mL	1.80
1.30	5.01×10^{-2} M						
		Second	35.20 mL	10.10	Second	26.40 mL	7.20

226

Molarity of the H_2SO_3 Solution
The molarity of the acid can be determined from either the first equivalence point or the second equivalence point; both give the same answer. In top equation, the molarity of the acid will be determined from the volume and molarity of the NaOH needed to reach the first equivalence point; noting that only one mole of NaOH is required to reach the first equivalence point. If the second equivalence point's volume is used (as in the bottom equation), the mole ratio would be 1 mole H_2SO_3 to 2 moles NaOH.

$$0.01760 \text{ L NaOH} \times \frac{0.400 \text{ mole NaOH}}{\text{L}} \times \frac{1 \text{ mole } H_2SO_3}{1 \text{ mole NaOH}} \times \frac{1}{0.0500 \text{ L } H_2SO_3} = 0.141 \text{ M } H_2SO_3$$

$$0.03520 \text{ L NaOH} \times \frac{0.400 \text{ mole NaOH}}{\text{L}} \times \frac{1 \text{ mole } H_2SO_3}{2 \text{ moles NaOH}} \times \frac{1}{0.0500 \text{ L } H_2SO_3} = 0.141 \text{ M } H_2SO_3$$

K_{a1} and K_{a2} Calculations
K_{a1} can be calculated from the first ½ equivalence point's pH by the equation $pH = pK_{a1}$.

pH at the first ½ EP = 1.80

$K_{a1} = 1.6 \times 10^{-2}$

The K_{a1} can also be determined using the ionization constant expression, because the initial molarity and pH ($[H^+]$) are known.

$$H_2SO_3 \, {}_{(aq)} \quad \rightleftharpoons \quad HSO_3^- \, {}_{(aq)} + H^+ \, {}_{(aq)}$$

	H_2SO_3	HSO_3^-	H^+
I	0.141	0	0
C	0.0501	0.0501	0.0501
E	0.0909	0.0501	0.0501

$$K_a = \frac{[COOH^-][H^+]}{[HCOOH]} = \frac{0.0501^2}{0.0909} = 2.76 \times 10^{-2}$$

It is also important to note that <u>the "x" term in the denominator of the ionization constant expression of polyprotic acids often cannot be taken to zero</u>; always check the percent ionization.

Both of these calculated K_{a1} values are in the same order of magnitude as the literature value of 1.7×10^{-2}.

Similarly, K_{a2} can be calculated from the second ½ equivalence point's pH by the equation $pH = pK_{a2}$.

pH at the second ½ EP = 7.20 $K_{a2} = 6.3 \times 10^{-8}$

The literature value of K_{a2} for H_2SO_3 is 6.4×10^{-8}.

Calculation of Second Equivalence Point

Calculate the pH at the 2nd EP using the following steps.
1. Write the acid-base reaction equation and write and complete its associated ICE table.

$$H_2SO_3 \text{ } _{(aq)} \quad + \quad 2 \text{ NaOH } _{(aq)} \quad \rightarrow \quad Na_2SO_3 \text{ } _{(aq)} + \quad 2 \text{ } H_2O \text{ } _{(l)}$$

I	(0.0500 L)(0.141 M)	(0.03520 L)(0.400 M)	0
C	−0.00705 moles	−0.0141 moles	+0.00705 moles
E	0 moles	0 moles	0.00705 moles

2. Calculate a new molarity for the NaCOOH. The new total volume is
    50.0 mL + 35.20 mL = 85.2 mL, or 0.0852 L.

$$H_2SO_3 \text{ } _{(aq)} \quad + \quad 2 \text{ NaOH } _{(aq)} \quad \rightarrow \quad Na_2SO_3 \text{ } _{(aq)} + \quad 2 \text{ } H_2O \text{ } _{(l)}$$

$$\frac{0.00705 \text{ moles}}{0.0852 \text{ L}}$$

$$= 0.0827 \text{ M}$$

3. Set up an ICE table based on the new molarity of the salt and the salt hydrolysis.
    Note that the reaction will be with only ONE mole of water.

$$SO_3^{2-} \text{ } _{(aq)} \quad + \quad H_2O \text{ } _{(l)} \quad \rightleftharpoons \quad HSO_3^- \text{ } _{(aq)} \quad + \quad OH^- \text{ } _{(aq)}$$

I	0.0827	0	0
C	−x	x	x
E	0.0827 − x	x	x

4. Set up the ionization constant expression for the above equilibrium values using
    K_{b2} (because OH^- is generated on the product side).

$$K_{b2} = \frac{[HSO_3^-][OH^-]}{[SO_3^{2-}]} \quad \text{or} \quad 1.5 \times 10^{-7} = \frac{[x][x]}{[0.0827 - x]}$$

5. Solve for x and calculate the pH.
    The x in the denominator can be taken to 0, because x is very small compared to
    0.0827. This gives x = 2.38 x 10^{-6} M.

$$x = [OH^-] = 1.11 \times 10^{-4} \text{ M}$$

$$pOH = -\log[OH^-] = 3.95$$

$$pH = 10.05$$

Equivalence Points in a Polyprotic Titration (a secondary check)
The mid-equivalence points in a polyprotic acid titration (i.e. not the last equivalence point) are quite difficult to calculate, as they are fourth or fifth degree polynomials of multiple components. However, they can be estimated quite accurately if the K_a values are known, and will be used as a secondary check for the pH values at the equivalence points.

The pH at a mid-equivalence point is the average of the two pKa values bracketing that equivalence point because here, the concentrations of both the conjugate acid of a species and the conjugate base of that same species will be equal.

$$\text{pH at } 1^{st} \text{ EP} = \tfrac{1}{2}\,(pK_{a1} + pK_{a2}) \qquad \text{(for diprotic and triprotic acids)}$$

$$\text{pH at } 2^{nd} \text{ EP} = \tfrac{1}{2}\,(pK_{a2} + pK_{a3}) \qquad \text{(for triprotic acids)}$$

Continuing the example of H_2SO_3, the pH at the first equivalence point in this diprotic acid titration can be confirmed.

$$\text{pH} = \tfrac{1}{2}\,(pK_{a1} + pK_{a2}) = \tfrac{1}{2}\,(1.80 + 7.20) = 4.50$$

This agrees well with the estimated pH of 4.60.

Post Equivalence Region in a Polyprotic Titration
In the post equivalence region, all of the acid is gone and the pH is dependent upon the amount and molarity of the base. The calculation of the pH uses the same method as a monoprotic acid.

The following summarizes the chemical reactions in weak acid-strong base titrations and weak base-strong acid titrations.

$HA \rightleftharpoons H^+ + A^-$	weak acid dissociation equation
$HA + NaOH \rightarrow NaA + H_2O$	weak acid + base reaction equation
$NaA \rightarrow Na^+ + A^-$	salt dissociation equation
$A^- + H_2O \rightleftharpoons HA + OH^-$	salt hydrolysis equation
$B + H_2O \rightleftharpoons BH^+ + OH^-$	weak base dissociation equation
$B + HCl \rightarrow BHCl$	acid + weak base reaction equation
$BHCl \rightarrow BH^+ + Cl^-$	salt dissociation equation
$BH^+ + H_2O \rightleftharpoons B + H_3O^+$	salt hydrolysis equation

Strong Base Titrated with a Strong Acid
For strong acid – strong base titrations, no buffer regions exist since a weak acid or weak base and its salt are not present. The equivalence point occurs at pH 7. Additionally, no K_a or K_b values are associated with strong acids and strong bases.

EXPERIMENT 16   ACIDS AND BASES: COMPLETE CURVE ANALYSIS

<u>Chemicals</u>

Potassium hydrogen phthalate	$KHC_8H_4O_{4\,(s)}$ (KHP)
Sodium hydroxide	$NaOH_{\,(aq)}$
Acetic acid	$HC_2H_3O_{2\,(aq)}$
Phosphoric acid	$H_3PO_{4\,(aq)}$
Hydrochloric acid	$HCl_{\,(aq)}$
Phenolphthalein indicator	
Bromocresol green indicator	

<u>Waste</u>

All waste and excess solutions can be disposed of down the drain.

<u>Procedure</u>

Set up and calibrate the digital pH meter according to the instructions in the Appendix. Unscrew the pH probe from the storage container (if applicable). Leave this top on during the calibration and all titrations.

<u>Standardization of the NaOH solution</u>

1. Normalize and then fill a buret using a small amount of the NaOH solution to be standardized.

2. Weigh out approximately 1.2 g (read to the nearest 0.0001 g) of the primary standard acid, $KHC_8H_4O_4$, KHP. KHP is a monoprotic acid with a molar mass of 204.23 g/mole.

3. Place the $KHC_8H_4O_4$ in a clean 125 mL Erlenmeyer flask. Add about 50 mL of deionized (DI) water and 3 to 4 drops of phenolphthalein.

4. Titrate until equivalence point has been reached, a faint pink color. During the titration, ensure all the KHP has dissolved. Wash the flask walls with DI water periodically to wash down any KHP or droplets of liquid that may have splashed up onto the sides of the flask.

5. Carefully touch the tip of the buret to the inside of the titration flask and make sure that this small amount of NaOH reaches the liquid in the flask.

6. Allow 30 seconds for drainage along the inner walls of the buret, then read and record the final NaOH volume.

7. Make a second determination in the same manner. They should agree within 5%. Between titrations rinse the Erlenmeyer flask once with tap water and then once with deionized water. It is not necessary to dry the flask.

<u>Titration of Acetic Acid with NaOH</u>

1. Using the auto dispenser, dispense 50.0 mL of the acetic acid unknown into a clean and dry 250 mL beaker. Add 3-4 drops of phenolphthalein. Swirl to mix. Do not add any water: this will change the starting pH value of the unknown acid.

2. Gently shake off any excess liquid on the pH electrode, blot with a paper towel, and then place the pH electrode in the beaker containing the acid solution. Record the pH. (0.00 mL NaOH added.) The electrode can be used as a stirring rod during the titration.

3. Add NaOH, a one mL at a time, reading and recording the pH and buret readings. When it appears that there is a more rapid rate of change of pH with volume added, the equivalence point should be close. At this point add NaOH 0.5 mL at a time. After the equivalence point is reached, continue the titration using 1 mL increments until the pH is above 12 for at least three readings.

Titration of Phosphoric Acid with NaOH
1. Using the auto dispenser, dispense 50.0 mL of the H_3PO_4 unknown into a well rinsed and dry 250 mL beaker. Since there is no single indicator that will correctly signal both endpoints in this titration, two different indicators must be used. Use 10 drops of bromocresol green for the titration to the first endpoint (the solution will be yellow to start, and then will be a very light blue at the equivalence point). After this first endpoint is reached, add 3 – 4 drops of phenolphthalein for the second endpoint, which is signaled by a change of color to pink/purple. Although H_3PO_4 has three equivalence points, the third equivalence point is so small that it is extremely difficult to determine, and will not be done in this laboratory. (Methyl red can be used in place of bromocresol green, but it is less stable. It is pink in its acidic form and a very light yellow in its basic form, when the phenolphthalein should be added.)

2. Gently shake off any liquid on the pH electrode, blot with a paper towel, place the electrode in the solution, and proceed to titrate as before, collecting and recording enough data to cover both equivalence regions and beyond (pH > 12).

Titration of HCl with NaOH
1. Obtain about 30 mL of the unknown HCl into a well rinsed and dry 100 mL beaker.

2. Normalize a 10.00 mL pipet with the acid solution and then deliver 10.00 mL of the acid to a clean and dry 250 mL beaker. Add about 100 mL of deionized water and 3-4 drops of phenolphthalein.

3. Gently shake off any liquid on the pH electrode, blot with a paper towel, place the electrode in the solution, and proceed to titrate as before, collecting enough data to cover the equivalence region and beyond (pH > 12).

4. Rinse the electrodes and replace them in their containers. Make sure the electrode and container remain upright for storage, and that there is plenty of solution in the container.

RINSE BURET WELL WITH DI H_2O,
FLIP, AND OPEN THE STOPCOCK

EXPERIMENT 16   ACIDS AND BASES: COMPLETE CURVE ANALYSIS

Lab Report for Acid Base Titrations

This lab report will be evaluated on data accuracy and on organization. The format must follow this order, including having the data and graph pages in the exact order specified. **The graphs and all prose must follow the directions provided (PC and MAC Excel instructions follow) explicitly, and must be independent from your partner's.**

Compilation, ancillary:
1. All pages, except the data pages, graphs, mathematical equations, conclusion, and error analysis must be typed. (The mathematical equations, conclusion and error analysis may be typed if time permits.) The graphs must be computer generated, in the exact format outlined in this set of instructions. The graphs in the discussion section of this experiment give examples of what the final graphs should look like.
2. No title page or report folder is necessary. The first page should have a right-justified heading containing your name, section, and date, each on a separate line.
3. All typed pages should have a centered page number in the footnote.
4. Staple the report in the upper left hand corner when complete.
5. Order the pages as follows: Title (centered and bolded), Purpose, Procedure, Chemicals Used, Waste Disposal, Chemical Reactions*, Mathematical Equations, Conclusion, Error Analysis, NaOH standardization page, HAc/NaOH data page, HAc/NaOH graph, H_3PO_4/NaOH data, H_3PO_4/NaOH graph, HCl/NaOH data, HCl/NaOH graph, the three "intermediate" graphs (see graphing instructions), in the order of experiments above. Be sure to complete the calculations on the back of each titration's data page.

   * There are eight chemical reactions: four are the molecular reactions that are performed in this experiment, and four are generic weak acid reactions.
   - To subscript, highlight the text and hit control and = together
   - For → and ⇌, find this in the equations tab (insert, equation, find the arrow or left over right harpoon in the pulldown menu of the symbols)

Conclusion
- Written in a technical manner, in complete sentences, using third person. ("It was determined")
- Discuss each titration curve separately. "Talk" through each curve in logical order, noting all significant titration points and titration parameters where applicable (important pH values and mL added; K_a values from both the ½ EP and ionization constant expression; molarities of the acid and base; etc.).
- Include units and insure significant figures are correct.

Error Analysis Questions (Use complete sentences and underline important data/answer.)
1. When standardizing the NaOH with KHP, if the KHP was overtitrated,
    a. How would this affect the molarity of NaOH?
    b. How would this affect the molarity of the acid unknowns?
2. How would the molarity of the acids and their calculated K_a values (where applicable) be affected if too much phenolphthalein (an acid) was added?
3. Could the acetic acid, phosphoric acid, and hydrochloric acid titrations of this experiment be performed without indicators? Explain.
4. When titrating the acid solutions, if the beaker was rinsed with DI water, but not dried,
    a. How would this affect the initial pH?
    b. How would this affect the volume of NaOH needed to reach the equivalence point?
5. For a weak base – strong acid titration, describe two methods that can be used to determine K_b (Listing the option $K_w = K_a K_b$ will not receive points for this question.)

Graphs
- Plot separate graphs for all titrations.
- Label axes, with units, if applicable; each graph's title must be very descriptive.
- Label each significant titration point or region correctly and neatly <u>where applicable</u> to the titration.
    o Initial pH
    o At ½ EP(s): mL(s) of NaOH added and pH
    o At EP(s): mL(s) of NaOH added and pH
    o Buffer region(s)
    o Post equivalence region
- Refer to the graphs in the discussion section for the proper format of the final graphs.

Plotting in Excel (General PC instructions):

1. Open Excel and enter data into two columns. The first column should be the Total mL NaOH added and the second column should be pH. To format both columns to two places after the decimal, highlight the columns, right click, select Format Cells, Number, and then enter 2.
2. Highlight both columns.
3. On the Insert tab, select Scatter in the Chart section, and then select Scatter with Smooth Lines and Markers from the pull down menu.
4. Right click inside the chart. Select Move Chart, then select New Sheet on the screen that pops up. Click OK.
5. Right click "Series 1" or "legend" or pH" and select Delete if these appear on the chart.
6. Right click the light blue or light gray border around the chart. Select Format Chart Area, Border, and then select No Line. Close the dialog box.
7. Select the Add Chart Element tab on the toolbar (this can also be done with the Layout tab).
   a. Add Chart Title. Increase the font size to 18 of the title when done.
   b. Add Axis Titles. Increase the font size to 14 of the axis labels when done.
   c. Double click on a value on the x axis (This menu can also be accessed by selecting More Axis or by right clicking on the numbers of the axis, selecting "Format Axis" and "Axis Options".). Set the Major unit to Fixed, 2.00, and the Minor unit to Fixed, 0.50. Select Outside for the Major tick mark type, and Inside for the Minor tick mark type. On the left toolbar of the Format Axis box, select "Number" and input "2" in the decimal places box. Close.
   d. Double click on a value on the y axis. Set the Major unit to Fixed, 1.00, and the Minor unit to Fixed, 0.10. Select Outside for the Major tick mark type, and Inside for the Minor tick mark type. On the left toolbar of the Format Axis box, select "Number" and input "2" in the decimal places box. Close.
8. Right click on the horizontal grid lines, and select Delete. Repeat for the vertical grid lines.
9. Right click on a number on the horizontal axis, and select Add Minor Grid Lines. Repeat for the vertical axis.
10. **Print out the graph, and manually determine the data for each significant point (see bullets on lab report directions for necessary points). This is the "intermediate" graph referred to in the compilation order section. <u>Write directly on these graphs to determine the values</u>.**
11. Right click on a number in the horizontal axis. Click Format Minor Grid Lines on the right pane and select No Line. Repeat for the vertical gridlines.
12. Click the Text Box icon in the Insert tab, and make a box near the point to be recorded. Fill in the text box with the point data (ex: 1st Equivalence Point: 11.20 mL, 3.78 pH). Size the text box accordingly. Click the Shapes icon on the Insert tab. Pick the arrow and draw an arrow from the text box to the place on the curve it depicts. Click the black color in the shape styles bar. Close the dialog box. Repeat for all applicable points. <u>Insert all text boxes describing specific points **below** the curve</u>.
13. To mark the buffer and post equivalence regions, select Shapes on the Insert tab and then select the bracket. Rotate the bracket using the green circle and then size accordingly. Click the black color in the shape styles bar. Repeat for all applicable regions. Add text boxes as appropriate to describe the regions. <u>These marks and corresponding text boxes should go **above** the curve</u>.
14. Redo the chart title (add mL and M of the acid and the M of the NaOH) after calculating values on the data sheet.
15. Print.

EXPERIMENT 16   ACIDS AND BASES: COMPLETE CURVE ANALYSIS

<u>Plotting in Excel (General Mac instructions):</u>
1. Open Excel and enter data into two columns. The first column should be the Total mL NaOH added and the second column should be pH. To format both columns to two places after the decimal, highlight the columns, right click, select Format Cells, Number, and then enter 2 in the area marked "Decimal places:". Click OK.
2. Highlight both columns.
3. Choose the "Charts" tab, click "Scatter" and then choose "Smooth Marked Scatter."
4. Right click inside the chart in the white space. Select Move Chart, and then select New Sheet on the screen that pops up. Click OK.
5. Right click "Series 1" or "pH" and select Delete.
6. Click on the tab "Chart Layout" on the tab next to "Charts".
   a. Click on "Chart Title". Select "Title Above Chart" and add a title. Increase the font size to 18 by selecting the box and changing the font size (or right click on the title and click format text)
   b. Add Axis Titles. To get a title for the x-axis, click on "Axis Titles" and select "Horizontal Axis Title", click "Title Below Axis", and enter the appropriate title. To get a title for the y-axis, select "Axis Titles", select "Vertical Axis Title", select "Rotated Title", and enter the appropriate title. Increase the font size font size to 14 of the axis labels when done.
   c. Double click on a value on the X Axis. Select "Scale" and set the Major unit to 2.00, and the Minor unit to 0.50. Select "Ticks" and select Outside for the Major tick mark type, and Inside for the Minor tick mark type. Click OK.
   d. Double click on a value on the Y Axis. Select "Scale" and set the Major unit to 1.00, and the Minor unit to 0.10. Select "Ticks" and select Outside for the Major tick mark type, and Inside for the Minor tick mark type. Click OK.
7. Click on one of the horizontal grid lines, and select Delete.
8. Right click a number on the horizontal axis; select Add Minor Grid Lines. Repeat for the vertical axis.
9. **Print out the graph, and manually determine the data for each significant point (see bullets on lab report directions for necessary points). This is the "intermediate" graph referred to in the compilation order section. <u>Write directly on these graphs to determine the values.</u>**
10. Click on the vertical and horizontal grid lines and select Delete.
11. To create a textbox, select Insert on the top tool bar, select Picture, then select Shape. Select "Rectangles", click on the farthest box to the left, and click on a space in the graph so the box will appear. Double click in the box and fill in the text box with the point data (ex. 1st Equivalence Point: 11.20 mL, 3.78 pH). Highlight the text and click on the tab "Home". Click on "Font Color" and select black. Change the font to Times New Roman, Font 12. Click on the tab "Format" and click on the down arrow of "Fill" and select No Fill. Click on the down arrow of "Line" and select No Line. Click on "Effects", click shadow and click on No Shadow. Size the text box accordingly.

    To create an arrow, select Insert on the top tool bar, select Picture, then select Shape. Click on "Lines and Connectors" and select the second picture on the top row on the left. Click on a space in the graph so the arrow will appear. Rotate the arrow so the pointer faces the curve at the point of interest. Click on the down arrow of "Line" on the toolbar and select black. Click on "Effects", click shadow and click on No Shadow. Size the arrow accordingly. Repeat for all applicable points. <u>Insert all text boxes describing specific points **below** the curve.</u>
12. To mark the buffer and post equivalence point regions, select Insert on the top tool bar, select Picture, then select Shape. Click on "Basic Shapes", select the "brace" (it should be the second to the last shape) and click on a space in the graph so the bracket will appear. Make the bracket bigger to fit the region being described and rotate the bracket using the green circle. Resize accordingly. Click on the tab "Format". Click on the down arrow of "Line" and select black. Click on "Effects", click shadow and click on No Shadow. Repeat for all applicable regions. Add text boxes as appropriate to describe the regions. <u>These marks and corresponding text boxes should go **above** the curve.</u>
13. Redo the chart title (add mL and M of the acid and the M of the NaOH) after calculating values on the data sheet.
14. Print.

LABORATORY REPORT SHEET: Standardization of NaOH

Name _______________________________________Date _____________

	Trial 1	Trial 2	Trial 3
Mass of $KHC_8H_4O_4$			
Initial buret reading			
Final buret reading			
Volume of NaOH added			

Calculation of NaOH molarity from Trial 1:

Calculation of NaOH molarity from Trial 2:

Calculation of NaOH molarity from Trial 3:

Calculation of selected trials ($< 5\%$):

Average NaOH molarity _____________________

LABORATORY REPORT SHEET: Acetic Acid ($HC_2H_3O_2$) and NaOH

Name ____Date _____________

TITRATION DATA

Acetic Acid ($HC_2H_3O_2$) and NaOH

Buret Reading, mL	Total mL NaOH added	pH		Buret Reading, mL	Total mL NaOH added	pH
*						

* This initial reading should be subtracted from subsequent buret readings (column 1) to obtain the total mL of NaOH added (column 2).

Acetic Acid ($HC_2H_3O_2$) and NaOH

NaOH molarity ________________

RECORD ALL VALUES WITH UNITS!

For this table, read the starting pH from the data table and the values at the EP and ½ EP from the graph.

Initial		Equivalence Point		½ Equivalence Point	
pH	$[H^+]$	mL	pH	mL	pH

Calculate the molarity of the acetic acid solution.

Calculate K_a from the ionization constant expression.

Calculate K_a from the ½ equivalence point.

Calculate the pH at equivalence point using ICE tables and the ionization constant expression.

LABORATORY REPORT SHEET: H_3PO_4 and NaOH

Name ____ Date ____________

TITRATION DATA

H_3PO_4 and NaOH

**RECORD ALL VALUES
WITH PROPER UNITS!**

Buret Reading, mL	Total mL NaOH added	pH		Buret Reading, mL	Total mL NaOH added	pH
*						

* This initial reading should be subtracted from subsequent buret readings (column 1) to obtain the total mL of NaOH added (column 2).

<u>H_3PO_4 and</u> __________ <u>M NaOH</u>

For this table, read the starting pH from the data table and the values at the EPs and ½ EPs from the graph.

Initial			Equivalence Points			½ Equivalence Points	
pH	$[H^+]$		mL	pH		mL	pH
		First			First		
		Second			Second		

Calculate the molarity of the phosphoric acid solution.

Calculate K_{a1} from the ionization constant expression.

Calculate K_a values from the
½ equivalence points:

K_{a1}	K_{a2}

Calculate the mL of NaOH to the 3rd EP using the volume and molarity of the acid.

Calculate the pH at the 3rd EP as follows. ($K_{a3} = 4.8 \times 10^{-13}$). Arrows must be correct.
a. Write the acid-base reaction equation and write and complete its associated ICE table.

b. The total volume is ________________ and the molarity of the salt ____________________.
c. Write the salt hydrolysis equation and complete the associated ICE table at the 3rd EP.

d. Write the ionization constant expression at the 3rd EP and calculate the pH. The quadratic
equation should be used, but not required if set up is correct.

LABORATORY REPORT SHEET: HCl and NaOH

Name ____Date ___________

TITRATION DATA

HCl and NaOH

RECORD ALL VALUES WITH PROPER UNITS!

Buret Reading, mL	Total mL NaOH added	pH		Buret Reading, mL	Total mL NaOH added	pH
*						

* This initial reading should be subtracted from subsequent buret readings (column 1) to obtain the total mL of NaOH added (column 2).

HCl and NaOH

NaOH molarity ________________

Read the starting pH from the data table and the values of the EP from the graph.

Initial		Equivalence Point	
pH	$[H^+]$	mL	pH

Calculate the molarity of hydrochloric acid.

<u>HOMEWORK EXERCISES</u>   Name _______________________________

This homework is six pages

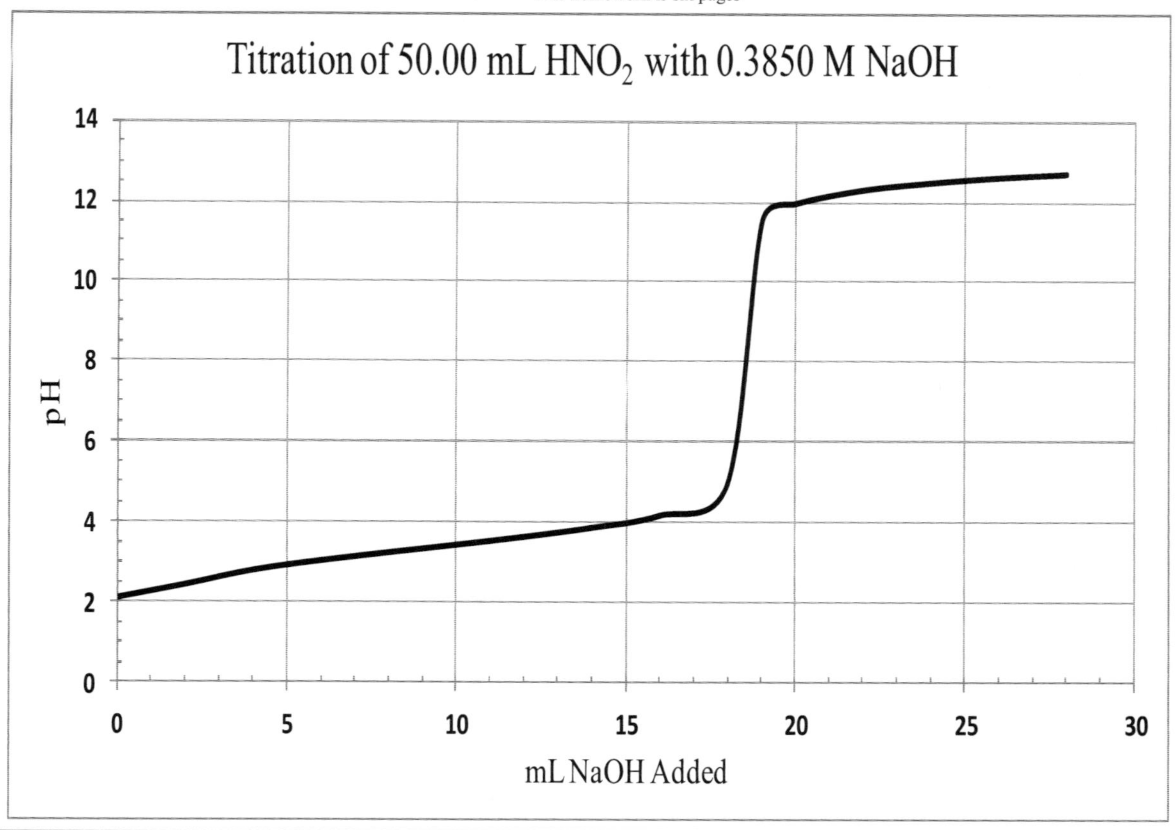

1. a.   Determine the following from the above graph:

Initial		Equivalence Point		½ Equivalence Point	
pH	$[H^+]$	mL	pH	mL	pH

   b.   Write the reactions with the appropriate arrows for the reaction between HNO_2 and NaOH:

   The HNO_2 with NaOH reaction equation:

   The salt dissociation equation:

   The salt hydrolysis equation:

   c.   What is the molarity of the HNO_2 solution? (Use the volume and molarity of the base.)

   d.   What is the value of K_a for HNO_2 calculated from the ½ equivalence point?

e. What is the value of K_a for HNO_2 calculated from the molarity and the initial pH?

f. At the equivalence point, what is the pH dependent upon?

g. Calculate the pH at the equivalence point using ICE tables and the ionization constant expression.

h. When 25.00 mL of NaOH have been added, what is the pH dependent upon?

i. When 25.00 mL of NaOH are added, what pH will the solution be?

2. The K_a value of formic acid, HCOOH, is 1.7×10^{-4}. 25.0 mL of 0.215 M formic acid are titrated with 0.250 M barium hydroxide.

    a. Write the chemical equation for this acid-base titration reaction.

    b. Calculate K_b _______________________________________

    c. Calculate the initial pH of the acid.

    d. Calculate the pH after 7.0 mL of barium hydroxide have been added.

    e. Calculate the volume of barium hydroxide required to reach the equivalence point.

f. Calculate the pH at the equivalence point.

g. Calculate the pH after 35.0 mL of barium hydroxide have been added.

h. Calculate the pH of the solution at the half equivalence point.

3. The pH of a 0.345 M weak base solution is 9.39. 50.0 mL of the weak base are titrated with 0.425 M HCl.
    a. Calculate the K_b of the weak base.

    b. Calculate the K_a of the weak base.

    c. Consult a Chemistry text and determine the identity of the weak base. ___________
    d. Calculate the volume of HCl required to reach the equivalence point.

e. Calculate the volume of HCl required to reach the ½ equivalence point.

f. Calculate K_b from the ½ equivalence point pH of 5.23.

g. Calculate the pH after 27.0 mL of HCl have been added.

h. Calculate the pH at the equivalence point.

i. Calculate the pH after 52.0 mL of the HCl have been added.

4. Complete the first table based on the following figure and use this data to complete the second table.

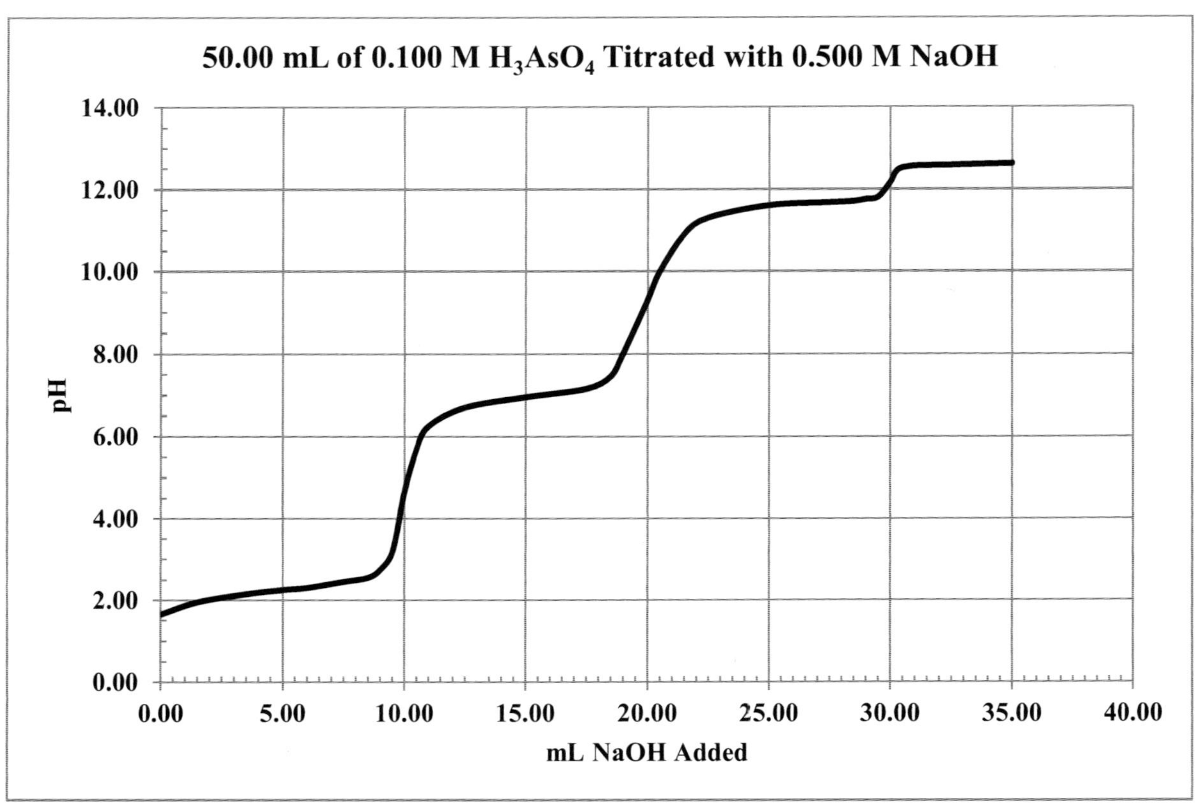

Initial			Equivalence Points			½ Equivalence Points	
pH	[H$^+$]		mL	pH		mL	pH
		First			First		
		Second			Second		
		Third			Third		

K_{a1}	
K_{a2}	
K_{a3}	
Check: pH at 1st EP from pK_a values $= \frac{1}{2}\,(pK_{a1} + pK_{a2})$:	
Check: pH at 2nd EP from pK_a values $= \frac{1}{2}\,(pK_{a2} + pK_{a3})$:	

These calculations are described in detail at the
following YouTube link: (43 minutes)
https://youtu.be/Iblk9CTeyF8   (the only numbers in the link are 9 and 8)

The exercises
given on the
following
pages do NOT
have to be
turned in.
They are for
practice.

Weak Acid – Strong Base Titration

75.0 mL of 0.250 M ascorbic acid ($HC_6H_7O_6$) are titrated with 0.200 M NaOH.

$K_{a, HC_6H_7O_6} = 8.0 \times 10^{-5}$ and $K_{b, HC_6H_7O_6} = 1.3 \times 10^{-10}$

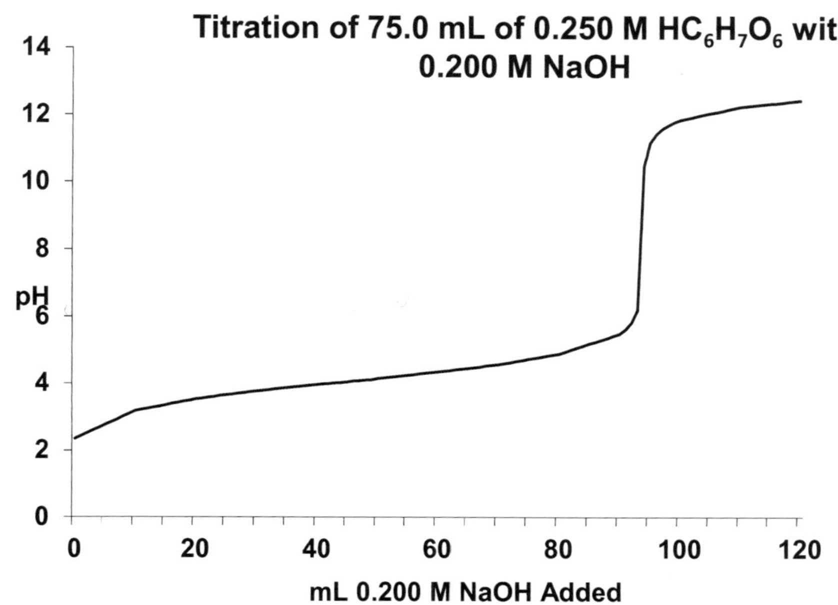

a.   What is the initial pH of the acid solution prior to starting the titration?

b.   Calculate the pH after the addition of 50.0 mL of 0.200 M NaOH.

c.   Calculate the volume of NaOH required to reach the equivalence point.

252

d.   Calculate the pH of the solution at the equivalence point.

e.   Calculate the pH after 120.0 mL of 0.200 M NaOH have been added.

Weak Base – Strong Acid Titration

100.0 mL of a 0.150 M methylamine solution are titrated with 0.250 M HCl. The initial pH is 11.91.

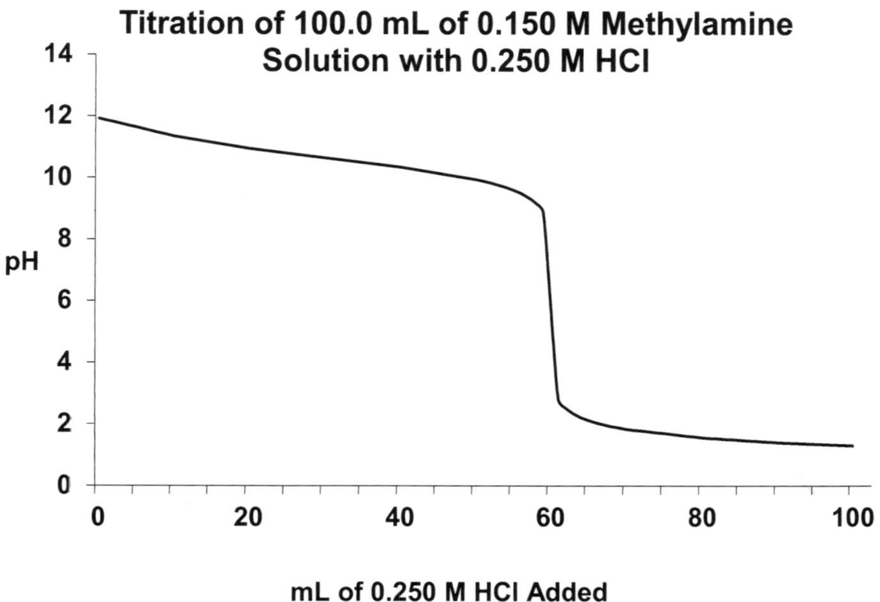

a.   Calculate the K_a and K_b values of the weak base.

b.   Calculate the pH after the addition of 40.0 mL of 0.250 M HCl.

c.   Calculate the volume of HCl required to reach the equivalence point.

d.   Calculate the pH of the solution at the equivalence point.

e.   Calculate the pH after 100.0 mL of 0.250 M HCl have been added.

EXPERIMENT 17
DETERMINATION OF THE SOLUBILITY PRODUCT CONSTANT OF $Cu(IO_3)_2$

The solubility product constant, K_{sp}, for the sparingly soluble substance, $Cu(IO_3)_2$ will be calculated. The concentration of Cu^{2+} will be determined through spectroscopy and the IO_3^- concentration will be determined through titration. The effect of common ions on solubility will also be investigated.

Key Chemical Reactions:

$$Cu(IO_3)_{2\ (s)} \ \rightleftharpoons\ Cu^{2+}_{\ (aq)} \ +\ 2\ IO_3^-{}_{\ (aq)}$$

$$Cu^{2+}_{\ (aq)} \ +\ 4\ NH_{3\ (aq)} \ \rightarrow\ [Cu(NH_3)_4]^{2+}_{\ (aq)}$$

$$IO_3^-{}_{\ (aq)} + 6\ H^+_{\ (aq)} + 8\ I^-_{\ (aq)} \ \rightarrow\ 3\ I_3^-{}_{\ (aq)} \ +\ 3\ H_2O_{\ (l)}$$

$$I_3^-{}_{\ (aq)} \ +\ 2\ S_2O_3^{2-}{}_{\ (aq)} \ \rightarrow\ 3\ I^-_{\ (aq)} \ +\ S_4O_6^{2-}{}_{\ (aq)}$$

Key Mathematical Equations:

$$K_{sp} = [Cu^{2+}][IO_3^-]^2$$

$$[Cu(NH_3)_4]^{2+}_{\ (aq)} = \frac{M oles\ of\ [Cu(NH_3)_4]^{2+}_{\ (aq)}}{0.01200\ L\ of\ solution}$$

$$L\ Na_2S_2O_3 \ x\ M\ Na_2S_2O_3 \ x\ \frac{1\ mole\ KIO_3}{6\ mole\ Na_2S_2O_3} \ x\ \frac{1}{L\ KIO_3} = [IO_3^-]$$

<u>Discussion</u>
Some ionic compounds are only slightly soluble in water, so that only a small amount of the solid dissociates into the ions that make it up. For example, silver carbonate, Ag_2CO_3, observes the following equilibrium:

$$Ag_2CO_{3\ (s)} \ \rightleftharpoons\ 2\ Ag^+_{\ (aq)} \ +\ CO_3^{2-}{}_{\ (aq)}$$

The equilibrium constant for this reaction is called the solubility product constant, K_{sp}. The equilibrium lies far to the left since only a small amount of the solid dissociates, so the values of K_{sp} are extremely small. The value of K_{sp} for an equilibrium reaction is the product of the molar concentrations of the ions, each raised to its stoichiometric coefficient. In the aqueous saturated solution of Ag_2CO_3, the relationship between K_{sp} and the concentrations is given by the equation:

$$K_{sp} = [Ag^+]^2[CO_3^{2-}]$$

255

Examples

The K_{sp} of Ag_2CO_3 at 25 °C is 8.10 x 10^{-12}. Calculate the molar solubility and solubility in g/L of Ag_2CO_3. An ICE table can be used to set up the equilibrium expression.

$$Ag_2CO_3 \, {}_{(s)} \quad \rightleftharpoons \quad 2 \, Ag^+ \, {}_{(aq)} \quad + \quad CO_3^{2-} \, {}_{(aq)}$$

Initial		0.0	0.0
Change	$- \, x$	$+ \, 2x$	$+ \, x$
Equilibrium		2x	x

$$K_{sp} = [Ag^+]^2[CO_3^{2-}]$$

$$8.10 \text{ x } 10^{-12} = [2x]^2[x]$$

$$x = 1.27 \text{ x } 10^{-4} \text{ M } = [Ag_2CO_3]$$

This is the molar solubility of Ag_2CO_3, with units of mol/L. The solubility in g/L is 0.035 g/L, calculated by multiplying the molar solubility by the molar mass of Ag_2CO_3 (275.7 g/mol). K_{sp} is a constant, and can only change if temperature changes.

When a common ion is present in solution, the solubility of the compound is affected. Calculate the solubility of Ag_2CO_3 in 0.0200 M Na_2CO_3 solution. The common ion in this situation is CO_3^{2-}, and is the "initial" value of the molarity of CO_3^{2-}. A new ICE table is constructed:

$$Ag_2CO_3 \, {}_{(s)} \quad \rightleftharpoons \quad 2 \, Ag^+ \, {}_{(aq)} \quad + \quad CO_3^{2-} \, {}_{(aq)}$$

Initial		0.0	0.0200
Change	$- \, x$	$+ \, 2x$	$+ \, x$
Equilibrium		2x	0.0200 + x

$$K_{sp} = [Ag^+]^2[CO_3^{2-}]$$

$$8.10 \text{ x } 10^{-12} = [2x]^2[0.0200 + x]$$

$$x = 1.01 \text{ x } 10^{-5} \text{ M } = [Ag_2CO_3]$$

> The "x" in the 0.0200 + x term can be taken to 0, since it is assumed that x is magnitudes of order smaller than 0.0200

Comparing the two solubilities, it can be seen that a common ion decreases the solubility of a compound. When a common ion is present, according to Le Châtelier's principle, the equilibrium will be shifted to the left, decreasing the concentration of the products.

In this experiment, three saturated $Cu(IO_3)_2$ solutions will be analyzed and the value of K_{sp} will be determined. The effect of a common ion in solution will also be investigated.

The Cu^{2+} concentrations will be determined using a spectrometer. Upon reaction with NH_3, the Cu^{2+} will be converted quantitatively to $[Cu(NH_3)_4]^{2+}$, which absorbs light intensely at 600 nm. This conversion is virtually complete ($K_f = 5.0$ x 10^{13}) so essentially all the copper is measured.

$$Cu^{2+} \, {}_{(aq)} \quad + \quad 4 \, NH_3 \, {}_{(aq)} \quad \rightarrow \quad [Cu(NH_3)_4]^{2+} \, {}_{(aq)}$$

EXPERIMENT 17   THE SOLUBILITY PRODUCT CONSTANT FOR $Cu(IO_3)_2$

Five solutions containing known concentrations of $[Cu(NH_3)_4]^{2+}$ will be prepared, their absorbance measured, and a calibration curve will be generated. The Cu^{2+} concentrations of the unknown solutions can then be obtained from this curve.

The IO_3^- concentrations will be determined by redox titration of the I_3^- formed when the IO_3^- reacts with I^-. In the first reaction, 3 moles of I_3^- are produced for every mole of IO_3^-. The I_3^-, which forms a dark blue complex with starch, can be reduced by titrating with standardized thiosulfate solution $(S_2O_3^{2-})$, as shown in the second reaction. Therefore, the number of moles of IO_3^- present in the solution will be equal to 1/6 of the number of moles of $S_2O_3^{2-}$ required to reach the endpoint of the titration. The following equations show the reaction and the reduction and the calculation of the IO_3^- concentration:

$$IO_3^-\ _{(aq)} + 6\,H^+\ _{(aq)} + 8\,I^-\ _{(aq)} \rightarrow 3\,I_3^-\ _{(aq)} + 3\,H_2O\ _{(l)}$$

$$I_3^-\ _{(aq)} + 2\,S_2O_3^{2-}\ _{(aq)} \rightarrow 3\,I^-\ _{(aq)} + S_4O_6^{2-}\ _{(aq)}$$

$$L\ Na_2S_2O_3 \text{ x } M\ Na_2S_2O_3 \text{ x } \frac{1\text{ mole }KIO_3}{6\text{ mole }Na_2S_2O_3} \text{ x } \frac{1}{L\ KIO_3} = [IO_3^-]$$

Chemicals

Ammonia (as NH_4OH)	$NH_3\ _{(aq)}$ (1M)
Copper (II) nitrate	$Cu(NO_3)_2\ _{(aq)}$ (0.0250 M)
Copper (II) iodate	$Cu(IO_3)_2\ _{(s)}$
Sodium citrate	$Na_3C_6H_5O_7\ _{(s)}$
Sulfuric acid	$H_2SO_4\ _{(aq)}$ (6 M)
Sodium thiosulfate	$Na_2S_2O_3\ _{(aq)}$ (0.0250 M – 0.0500 M)
Iodine	$I_2\ _{(aq)}$
Potassium iodide	$KI\ _{(s)}$
Starch indicator	

Waste

All waste, except for excess sodium thiosulfate solution, should be collected and disposed in the waste container. Excess sodium thiosulfate can be disposed down the drain.

Procedure

Construction of the calibration curve for Cu^{2+}

1. Normalize and fill two burets, one with 0.0250 M $Cu(NO_3)_2$ solution and one with deionized water.

2. Using five clean, dry, and labeled test tubes or volumetric flasks, prepare and mix the solutions listed in Table 1 on the Laboratory Report Sheet. Use a 2.00 mL glass pipette to add the 1.00 M NH_3 solution. Stopper each test tube or flask and mix by gently inverting the test tube a number of times. If only one stopper is available, either dry it completely before using it on another test tube or use a stirring rod.

3. Calibrate the spectrometer with the given instructions. The same spectrometer cell (cuvette) should be used for all samples, and it should be normalized prior to each reading.

4. Normalize the cuvette with one of the five solutions. Fill the cuvette to about a ¼ to ½ inch from the top, insert into the spectrometer and read the absorbance at 600 nm.

5. Normalize, fill and read (at 600 nm) each of the remaining solutions. Record the results in Table 1. Calculate the moles of Cu^{2+} and $[Cu(NH_3)_4]^{2+}$ using the data in Table 1 and the following equation to establish the Cu^{2+} to $[Cu(NH_3)_4]^{2+}$ mole ratio:

$$Cu^{2+}_{(aq)} \;+\; 4\,NH_{3\,(aq)} \;\rightarrow\; [Cu(NH_3)_4]^{2+}_{(aq)}$$

6. Calculate the concentrations of $[Cu(NH_3)_4]^{2+}$ present using the following formula (each solution contains 12.00 mL per Table 1):

$$[Cu(NH_3)_4]^{2+}_{(aq)} = \frac{Moles\ of\ [Cu(NH_3)_4]^{2+}_{(aq)}}{0.01200\ L\ of\ solution}$$

7. Plot a curve of the line as follows:
   1. Minimize the spectrometry program. **DO NOT EXIT THE PROGRAM**.
   2. Go to Start, All Programs, Microsoft Office, Excel.
   3. Enter "Complex Molarity" in the first column and "Absorbance" in the second column.
   4. Enter the data from Table 1, including the point (0,0).
   5. Highlight the data, click on the Data tab and click on the Sort A to Z icon.
   6. Highlight the data, right click, and choose Format Cells. Click the Number tab and select Number from the Category list. Set Decimal Places to 3. Click OK.
   7. Highlight the data, click the Insert tab. Click the arrow under Scatter, choose the Scatter with smooth lines and markers. Move the chart underneath the data.
   8. Right click on one of the data points, click Add Trendline and select Linear. Check the boxes to "Display Equation on chart" and "Display R-squared value on chart". Close.
   9. Under Chart Tools, click the Layout tab.
      a. Click Chart Title, click Above Chart. Add a title.
      b. Click the Axis Titles icon. Click Primary Horizontal Axis Title, click Title Below Axis. Add the x axis title.
      c. Click the Axis Titles icon. Click Primary Vertical Axis Title, click on Rotated Title. Add the y axis title.
   10. Delete the Legend.
   11. Right click on any horizontal line in the chart. Select Delete.
   12. Move the box containing the equation of the line and R^2 value to any open area below the line. Also record these values on the Laboratory Report Sheet.
   13. If capabilities in the lab allow, print one graph for each partner. If not, show the graph (on the computer screen) to the instructor.
   14. Exit, and DO **NOT** SAVE.

Analysis of the Solutions
The three saturated unknown solutions (two with common ions) have already been prepared and allowed to reach equilibrium.

Analysis of [Cu^{2+}]
1. Dispense 10.00 mL samples of each of the three unknown solutions from the burets directly into dry and labeled volumetric flasks or test tubes and add 2.00 mL of 1.00 M NH$_3$ to each. Mix, making sure no cloudiness exists.

2. Normalize and then refill the cuvette with the first solution. Read and record the absorbance at 600 nm. Repeat for the remaining two solutions.

3. Use equation of the line from the calibration graph and the absorbance values to determine the concentrations of [Cu(NH$_3$)$_4$]$^{2+}$ present in each solution. Record the [Cu^{2+}] in all the solutions. Since the calibration graph was also based upon 12.00 mL of solution, [Cu^{2+}] = [Cu(NH$_3$)$_4$]$^{2+}$.

4. Turn off the spectrometer's lamp and exit the program. Shut down the computer. Collect all waste and excess solutions and dispose of them in the waste container.

Analysis of [IO$_3^-$]
1. Normalize a 50 mL buret and fill it with the Na$_2$S$_2$O$_3$ solution.

2. Dispense 10.00 mL of the saturated Cu(IO$_3$)$_2$ solution (from the burets) into a 125 mL flask. Add about 50 mL of DI water, one disposable pipetful of 6 M H$_2$SO$_4$, one rounded scoop of KI, and two rounded scoops of sodium citrate (to prevent the precipitation of CuI).

3. Titrate the copper iodate solution with the Na$_2$S$_2$O$_3$ solution until the brownish-yellowish color is gone and a light yellow appears. When a light yellow color appears, add about 10 drops of starch indicator solution. (Do not wait until the solution is clear!) The solution will turn dark blue/black. Continue the titration until the color changes from the intense blue starch–I$_3^-$ complex to "water white". An extremely pale blue color may be observed due to the presence of Cu^{2+}. The clear solution may turn back to an intense blue solution within 60 seconds. The correct titration reading is the first time the solution turns clear.

4. Repeat steps 1-3 for the remaining saturated solutions.

5. Using the volume and molarity of Na$_2$S$_2$O$_3$, calculate the moles of S$_2$O$_3^{2-}$ present and the concentration of IO$_3^-$ in the original solution.

LABORATORY REPORT SHEET

Name____Date___________

Table 1.   *Absorbance of Calibration Solutions*

Test tube	mL of 0.0250 M $Cu(NO_3)_2$	mL of 1.00 M NH_3	mL of H_2O	Moles of Cu^{2+}	Moles of $[Cu(NH_3)_4]^{2+}$	Molarity of $[Cu(NH_3)_4]^{2+}$	Absorbance
1	10.00	2.00	0.00				
2	8.00	2.00	2.00				
3	6.00	2.00	4.00				
4	4.00	2.00	6.00				
5	2.00	2.00	8.00				

Plot Absorbance vs. Molarity of $[Cu(NH_3)_4]^{2+}$ to obtain a calibration graph.

Equation of the line: ____

R^2 value: _______________________________

Continued on back

LABORATORY REPORT SHEET

RECORD ALL VALUES WITH PROPER UNITS!

Name________________________________ Date________________

Table 2.   Solution Data and Determination of K_{sp}

		Saturated $Cu(IO_3)_2$ in DI H_2O	Saturated $Cu(IO_3)_2$ in $Cu(NO_3)_2$	Saturated $Cu(IO_3)_2$ in KIO_3
	Molarity of common ion solution	None		
$[Cu^{2+}]$ Solution Data	Absorbance			
	Molarity of $[Cu(NH_3)_4]^{2+}$			
	$[Cu^{2+}]$ in original solution			
$[IO_3^-]$ Solution Data	Molarity of $Na_2S_2O_3$			
	Initial Buret Reading			
	Final Buret Reading			
	mL of $S_2O_3^{2-}$ used			
	mL of $Cu(IO_3)_2$ solution titrated			
	$[IO_3^-]$ in $Cu(IO_3)_2$ in DI H_2O *show calculation*			
	$[IO_3^-]$ in $Cu(IO_3)_2$ in $Cu(NO_3)_2$ *show calculation*			
	$[IO_3^-]$ in $Cu(IO_3)_2$ in KIO_3 *show calculation*			
K_{sp} Data	Chemical reaction of equilibrium for $Cu(IO_3)_2$			
	K_{sp} equation			
	K_{sp}			
	Average K_{sp}			

K_{sp} Error Analysis

1. For each of the following two errors, **complete the sentence**, stating (1) what component of the K_{sp} equation (Cu^{2+}, IO_3^-, or both) is affected, (2) how that component is affected, and (3) exactly how the error affects the experimentally determined K_{sp}.

    If the copper (II) iodate solution was over titrated,

    If there were scratches or fingerprints on the cuvette,

2. Using the experimental data, complete the table and answer the following questions.

	Solution 1 Cu(IO$_3$)$_2$ in DI H$_2$O		Solution 2 Cu(IO$_3$)$_2$ in Cu(NO$_3$)$_2$		Solution 3 Cu(IO$_3$)$_2$ in KIO$_3$	
Molarity and identity of common ion in solution						
	$[Cu^{2+}]$	$[IO_3^-]$	$[Cu^{2+}]$	$[IO_3^-]$	$[Cu^{2+}]$	$[IO_3^-]$
Experimentally determined concentrations						
K_{sp}						
Generic ICE Table for each solution (Set up only)	$Cu(IO_3)_{2\ (s)} \rightleftharpoons Cu^{2+}_{(aq)} + 2\ IO_3^-{}_{(aq)}$ I C E		$Cu(IO_3)_{2\ (s)} \rightleftharpoons Cu^{2+}_{(aq)} + 2\ IO_3^-{}_{(aq)}$ I C E		$Cu(IO_3)_{2\ (s)} \rightleftharpoons Cu^{2+}_{(aq)} + 2\ IO_3^-{}_{(aq)}$ I C E	

In Solution 2, $[Cu^{2+}]$ should be $\approx$ molarity of the common ion in solution, and in Solution 3, $[IO_3^-]$ should be $\approx$ molarity of the common ion in solution. Does this seem to be a good approximation? When a common ion is present, why can the "C" term of the common ion from the ICE table usually be ignored?

Continued on back

3. In the determination of the solubility for copper (II) iodate, what common ion of the same concentration (for example, 0.010 M Cu^{2+} or 0.010 M IO$_3^-$) would have a greater effect on reducing the solubility of Cu(IO$_3$)$_2$? Why?

4. A determination of K$_{sp}$ can be done by measuring only [Cu^{2+}] or [IO$_3^-$]. This statement is valid for only what situation?

5. Calculate the solubility of copper (II) iodate in DI H$_2$O in g/L.

6. Qualitatively compare the experimentally determined K$_{sp}$ to that of the literature value, 1.4 x 10^{-7}.

Name ________________________________

HOMEWORK EXERCISES

1. In this experiment, solutions will be analyzed to determine the Cu^{2+} concentration present.

   a. Write the chemical reaction that the copper undergoes to make a complex ion.

   b. Why is it important to convert the copper to this complex ion?

2. Show the net reaction of the following balanced equations and calculate the ratio of the moles of IO_3^- present to the moles of $S_2O_3^{2-}$ required to reach the endpoint of the titration.

$$IO_3^-\,{}_{(aq)} + 6\,H^+{}_{(aq)} + 8\,I^-{}_{(aq)} \;\rightarrow\; 3\,I_3^-{}_{(aq)} \;+\; 3\,H_2O\,{}_{(l)}$$

$$I_3^-{}_{(aq)} \;+\; 2\,S_2O_3^{2-}{}_{(aq)} \;\rightarrow\; 3\,I^-{}_{(aq)} \;+\; S_4O_6^{2-}{}_{(aq)}$$

3. In an experiment, 10.00 mL of Cu(IO$_3$)$_2$ were mixed with a scoop of KI, a scoop of sodium citrate, and small amount of sulfuric acid. 15.22 mL of 0.0473 M Na$_2$S$_2$O$_3$ were required to titrate to the endpoint. What is the concentration of $[IO_3^-]$ in the original solution?

Continued on back

4. In an experiment, a saturated solution saturated of Cr(OH)$_3$ was analyzed and the total Cr^{3+} concentration was found to be 3.25 x 10^{-8} M and the total [OH$^-$] was determined to be 9.74 x 10^{-8} M.

    a. Write the K$_{sp}$ equation for Cr(OH)$_3$. _______________________________________

    b. Calculate the value of K$_{sp}$ for Cr(OH)$_3$.

5. Using the value of K$_{sp}$ calculated in (4) above, set up and ICE table and calculate the following:
    a. Molar solubility and solubility in g/L of Cr(OH)$_3$ in pure water.

    b. Molar solubility and solubility in g/L of Cr(OH)$_3$ in 0.0502 M Cr(NO$_3$)$_3$.

    c. Molar solubility and solubility in g/L of Cr(OH)$_3$ in 0.0752 M NaOH.

6. Write the K$_{sp}$ equation for Ca$_3$(PO$_4$)$_2$. _______________________________________

EXPERIMENT 18
ELECTROCHEMISTRY AND THERMODYNAMICS

The E of different cell combinations will be measured, and G and K calculated. A voltaic cell system will be heated and E will be measured at different temperatures. G, H, and S will be calculated and general conclusions will be made about the spontaneity of reactions based on thermodynamic data. Finally, E will be determined for different concentrations of solutions, making general conclusions about the Nernst Equation.

Key Chemical Reactions:

$$Ni_{(s)} + Cu^{2+}_{(aq)} \rightarrow Cu_{(s)} + Ni^{2+}_{(aq)} \qquad Ni\,(s)\ |\ Ni^{2+}\,(0.10M)\ \|\ Cu^{2+}\,(0.10\ M)\ |\ Cu\,(s)$$

Key Mathematical Equations:

$$E^{\circ}_{cell} = E^{\circ}_{cathode} - E^{\circ}_{anode} \qquad \Delta G = -nFE \qquad G = -RT \ln K$$

$$\Delta G = \Delta H - T\Delta S \qquad E = E^{0} - \frac{RT}{nF}\ln Q$$

<u>Discussion</u>

Electrochemistry deals with the relationships and changes between chemical energy and electrical energy. The chemical reactions in electrochemistry are redox, or oxidation-reduction reactions, involving the transfer of electrons between substances. Electrochemical cells are set up as either voltaic (galvanic) cells, where chemical reactions produce electricity, or electrolytic cells, where electricity is used to produce a chemical change.

Both voltaic and electrolytic cells are made up of two half cells: the anode, the electrode where oxidation occurs, and the cathode, the electrode where reduction occurs. Each half cell contains an electrode partially immersed in an aqueous solution of the electrode metal. This is shown in Figure 18.1 for a voltaic cell.

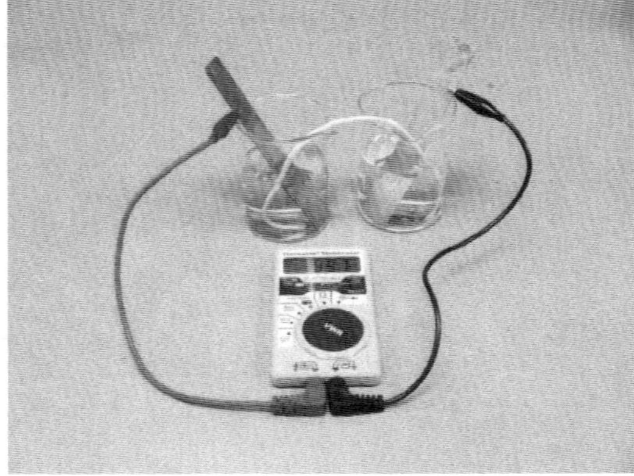

Figure 18.1. Voltaic Cell Set Up

In the above figure, the two half cells are connected by a salt bridge, so that the ions can move between the cells and electrical equilibrium is achieved. An inert salt such as

KNO_3, one that will not react with either solution or electrode, is used in the salt bridge. The chemical changes represented in this voltaic cell are expressed by the following half reactions:

$$\text{Oxidation (anode):} \qquad Zn_{(s)} \rightarrow Zn^{2+}_{(aq)} + 2\,e^-$$

$$\text{Reduction (cathode):} \qquad Cu^{2+}_{(aq)} + 2\,e^- \rightarrow Cu_{(s)}$$

There is standard annotation for combining these half reactions in a cell, called a cell diagram, written:

$$Zn\,(s) \mid Zn^{2+}\,(1\ M) \parallel Cu^{2+}\,(1\ M) \mid Cu\,(s)$$

$$\underbrace{\text{Oxidation (Anode)}} \quad \text{Salt Bridge} \quad \underbrace{\text{Reduction (Cathode)}}$$

The left part of the diagram is the anode half reaction, and the right part of the diagram is the cathode half reaction. The double line represents the salt bridge and the single lines show that there is a phase boundary. The concentration of aqueous solutions is written in parenthesis after the ion. If no concentration is noted for an aqueous solution or gas, it is assumed to be 1 M or 1 atm, respectively.

Again referring to Figure 18.1, electrons flow from the zinc anode (right cell in picture) through the voltmeter to the copper cathode. As the zinc electrode loses electrons, Zn^{2+} ions build up in the solution. At the cathode in the other half cell, Cu^{2+} ions from the $Cu(NO_3)_2$ solution combine with the electrons and plate out as Cu onto the copper electrode. The salt bridge provides electrical balance by transferring anions (NO_3^-) to the Zn cell and cations (K^+) to the Cu cell.

The voltage for each half reaction that is written as a reduction reaction is called the standard reduction potential or $E°$. The ° denotes standard state conditions, 298 K and 1 M for aqueous solutions or 1 atm for gases. Because $E°$ is an intensive property, it does not depend on the amount of material present. If the reaction is reversed, the sign of $E°$ changes. Table 18.1 gives Standard Reduction Potentials for various reductions.

Table 18.1. Standard Reduction Potentials

Reduction Half Reaction	$E°$ (V)
$Au^{3+}_{(aq)} + 3\,e^- \rightarrow Au_{(s)}$	+ 1.50
$Hg_2^{2+}_{(aq)} + 2\,e^- \rightarrow 2\,Hg_{(l)}$	+ 0.92
$Ag^+_{(aq)} + e^- \rightarrow Ag_{(s)}$	+ 0.80
$Cu^{2+}_{(aq)} + 2\,e^- \rightarrow Cu_{(s)}$	+ 0.34
$2\,H^+_{(aq)} + 2\,e^- \rightarrow H_{2\,(g)}$	0.00
$Pb^{2+}_{(aq)} + 2\,e^- \rightarrow Pb_{(s)}$	− 0.13
$Sn^{2+}_{(aq)} + 2\,e^- \rightarrow Sn_{(s)}$	− 0.14
$Ni^{2+}_{(aq)} + 2\,e^- \rightarrow Ni_{(s)}$	− 0.25
$Fe^{2+}_{(aq)} + 2\,e^- \rightarrow Fe_{(s)}$	− 0.44
$Cr^{3+}_{(aq)} + 3\,e^- \rightarrow Cr_{(s)}$	− 0.74
$Zn^{2+}_{(aq)} + 2\,e^- \rightarrow Zn_{(s)}$	− 0.76
$Mn^{2+}_{(aq)} + 2\,e^- \rightarrow Mn_{(s)}$	− 1.18
$Al^{3+}_{(aq)} + 3\,e^- \rightarrow Al_{(s)}$	− 1.66
$Mg^{2+}_{(aq)} + 2\,e^- \rightarrow Mg_{(s)}$	− 2.37
$Na^+_{(aq)} + e^- \rightarrow Na_{(s)}$	− 2.71
$Ca^{2+}_{(aq)} + 2\,e^- \rightarrow Ca_{(s)}$	− 2.87

To calculate the standard emf of a cell ($E°_{cell}$), first determine if the cell is voltaic or electrolytic from the problem statement. If the cell is voltaic, the $E°_{cell}$ needs to be a positive value. If the cell is electrolytic, the $E°_{cell}$ needs to be a negative value. Use the following equation, adding the oxidation (anode) and reduction (cathode) potentials together, *reversing* the sign of the oxidation reaction's potential:

$$E°_{cell} = E°_{oxidation} + E°_{reduction} \quad OR \quad E°_{cell} = E°_{anode} + E°_{cathode}$$

For the voltaic cell ($E°_{cell}$ must be +)   Zn (s) $|$ Zn^{2+} (1 M) $\|$ Cu^{2+} (1 M) $|$ Cu (s), the standard reduction potential is as follows (addition of the −oxidation and reduction $E°$):

$$E°_{cell} = 0.76 \text{ V} + 0.34 \text{ V} = 1.10 \text{ V}$$

Connections to Thermodynamics
The thermodynamic quantity, ΔG, called the change in Gibbs free energy, is the energy available to do work. It is defined using the following formula:

$$\Delta G = -nFE$$

where
    n is moles of electrons transferred (using the cell's net balanced redox reaction)
    F is Faraday's Constant (1 F = 96,500 J/V-mol e^- = 9.647×10^4 C/mol e^-)
    E is the measured cell voltage

The sign of ΔG determines the spontaneity of the reaction: spontaneous reactions have a negative ΔG, while non-spontaneous reactions have a positive ΔG.

ΔG is also related to the equilibrium constant of a reaction, K, with the following equation:
$$\Delta G = -RT \ln K$$
where
    R is the ideal gas constant, 8.314 J/K-mol
    T is the temperature in K
    K is the equilibrium constant
        K is extremely large (equilibrium lies with the products) for voltaic cells ($-\Delta G$)
        K is extremely small (equilibrium lies with the reactants) for electrolytic cells ($+\Delta G$)

It should be noted in this equation that since the units of ΔG are kJ/mol, this value should be converted to J/mol so that units cancel.

ΔG is also related to the enthalpy and entropy of the system with the following equation:

$$\Delta G = \Delta H - T\Delta S$$

Graphically, this is a line with the form y = b + mx. Plotting ΔG vs T, the slope is $-\Delta S$, and the y-intercept is ΔH.

These three identities are summarized as follows:

$$\Delta G = -nFE = -RT \ln K = \Delta H - T\Delta S$$

The following two tables summarize the determination of the spontaneity of a reaction.

Table 18.2. Spontaneity of Reactions based on E (and Relation to ΔG and K)

E	ΔG $(= -n\,F\,E)$	K $(\Delta G = -RT \ln K)$	Spontaneity
+	−	>1	• Reaction is spontaneous in the direction written • Voltaic cells
−	+	<1	• Reaction is not spontaneous in the direction written (spontaneous in the reverse direction) • Electrolytic cells
0	0	=1	

Table 18.3. Spontaneity of Reactions based on ΔH, ΔS, and T

$\Delta G = \Delta H - T\Delta S$			
ΔH	ΔS	ΔG	Spontaneity
+	+	+ or −	Spontaneous at HIGH temperatures only
+	−	+	Not spontaneous in the direction written
−	+	−	Spontaneous in the direction written
−	−	+ or −	Spontaneous at LOW temperatures only

<u>Effect of Varying Concentration on E</u>
Varying the concentration of the solutions in a cell changes the emf of a cell. The Nernst Equation is used to calculate the new E.

$$E = E^0 - \frac{RT}{nF} \ln Q$$

Where Q is the reaction quotient, $\frac{[product]^y}{[reactant]^x}$, and x and y are the stoichiometric multipliers in the balanced chemical equation. Note that solids are not included in the equation.

At 298 K (only), this equation becomes

$$E = E^0 - \frac{0.0257\ V}{n} \ln Q$$

If the temperature is not at 298 K, determine a new "E°" (here called E') with the same concentrations, and then complete the Nernst Equation.

Example:

The following data were obtained from an Ag-Cu voltaic cell system. The electrodes were immersed in 1 M $AgNO_3$ and 1 M $Cu(NO_3)_2$, respectively.

T	E
298 K	0.460 V
310 K	0.452 V
320 K	0.444 V
330 K	0.435 V

a. For the reaction:
   - Calculate $E°$
   - Write the two the half-cell reactions and identify the anode and cathode
   - Write the balanced net ionic equation
   - Draw the cell diagram for the reaction run as a voltaic cell

Since this is a voltaic cell, the E must be positive. Using the emf table, to make the E positive, the copper must be oxidized (flip the equation!) and the silver must be reduced.

$E°$:   0.80 V – 0.34 V = 0.46 V

Half cell reaction for Cu: $Cu_{(s)} \rightarrow Cu^{2+}_{(aq)} + 2\,e^-$
                                           Anode (oxidation) = copper

Half cell reaction for Ag: $Ag^+_{(aq)} + e^- \rightarrow Ag_{(s)}$
                                           Cathode (reduction) = silver

Balanced net ionic equation: $Cu_{(s)} + 2\,Ag^+_{(aq)} \rightarrow Cu^{2+}_{(aq)} + 2\,Ag_{(s)}$

Cell diagram: $Cu\,(s)\,|\,Cu^{2+}\,(1\ M)\,\|\,Ag^+\,(1\ M)\,|\,Ag\,(s)$

b. Calculate G and K at each temperature. The calculations at 298 K are shown.

$\Delta G\ (= -\,n\,F\,E) = -\,(2\text{ moles }e^-)(96{,}500\text{ J/V-mol }e^-)(0.460\text{ V})(1\text{ kJ/1000J}) = -\,88.8\text{ kJ/mol}$

$K\ (\Delta G = -RT\ln K)$: $-88.8\ \times\ 10^3\text{ J/mol} = -(8.314\text{ J/mol-K})(298\text{ K})(\ln K)$; $K = 3.68\times 10^{15}$

T (K)	E (V)	$\Delta G\ (= -\,n\,F\,E)$	$K\ (\Delta G = -RT\ln K)$
298	0.460	– 88.8 kJ/mol	3.68×10^{15}
310	0.452	– 87.2 kJ/mol	4.94×10^{14}
320	0.444	– 85.7 kJ/mol	9.76×10^{13}
330	0.435	– 84.0 kJ/mol	1.98×10^{13}

c. From the data in (b), determine ΔS and ΔH.

Plotting ΔG vs T gives a straight line with the form y = b + mx. This is analogous to the linear equation $\Delta G = \Delta H - T\Delta S$.

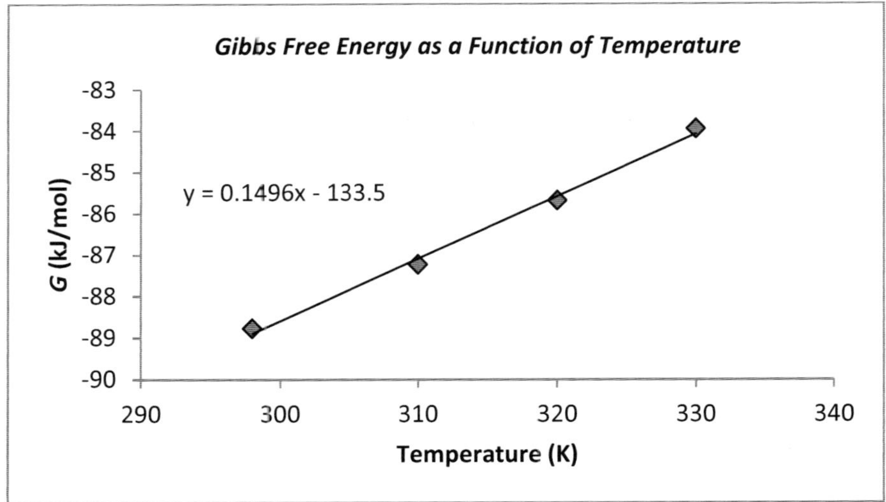

Using the data, the slope of the line is $-\Delta S$. Since the slope was determined to be +0.1496, this gives a ΔS of -0.1496 kJ/mol-K, or -149.6 J/mol-K. The y intercept, ΔH, is -133.5 kJ/mol.

d. Make general statements about the spontaneity of the reaction based on E, ΔG, and ΔH, and the position of equilibrium based on K.

Since E is positive, the reaction is spontaneous.
Since G is negative, the reaction is spontaneous.
Since H is negative, S is negative, and G is negative, this reaction is spontaneous in the direction written.
Since K is very large (>>>1), the equilibrium lies with the products.

e. If the concentration of the $AgNO_3$ is 0.1 M, calculate the E and make predictions about the values of ΔG, ΔH, and K.

The Nernst Equation is $E = E^0 - \dfrac{RT}{nF}\ln Q$, where Q is the reaction quotient, $\dfrac{[product]^y}{[reactant]^x}$.

$$E = 0.460\ V - \frac{(8.314\frac{J}{mol-K})(298\ K)}{(2\ mol\ e^-)(96500\frac{J}{V-mol\ e-})}\ln\frac{1}{(0.1)^2} = 0.401\ V$$

Since E is positive, the reaction is spontaneous. Therefore, ΔG and ΔH must be negative, and K must be very large.

<u>Chemicals</u>

Copper	Cu (copper colored electrode)
Copper nitrate	$Cu(NO_3)_2$
Lead	Pb (heavy, long, dark electrode; bends easily)
Lead nitrate	$Pb(NO_3)_2$
Potassium nitrate or sodium nitrate	KNO_3 or $NaNO_3$
Nickel	Ni (electrode marked Ni at top)
Nickel nitrate	$Ni(NO_3)_2$
Zinc	Zn (short, light, shiny electrode)
Zinc nitrate	$Zn(NO_3)_2$

<u>Waste</u>   <u>(DO THIS AT THE END OF THE EXPERMENT: SOME SOLUTIONS WILL BE USED MORE THAN ONCE!)</u>

All liquid waste will be recycled to the original container or as directed by the instructor. Used salt bridges should be disposed in the trash.

<u>Procedure</u>

Part 1

1. Sand off any oxidation from the copper, nickel, lead, and zinc electrodes and wipe with a paper towel. Dip in 1 M HCl and then dry thoroughly.

2. Fill four beakers ¾ full with the 0.1 M metal solutions of $Cu(NO_3)_2$, $Ni(NO_3)_2$, $Pb(NO_3)_2$, $Zn(NO_3)_2$.

3. Place the dry electrodes into their respective solutions in the four beakers.

4. Remove six 6" pieces of string (salt bridges) that have been immersed in 0.1 M KNO_3 or 0.1 M $NaNO_3$. Gently blot them on a paper towel to remove excess solution.

5. Insert the salt bridges as shown to connect pairs of cells, making sure they do not touch the electrodes or each other.

6. Attach the black plug to the anode and the red plug to the cathode of a cell system. Set the multimeter to DC and record the voltage. Repeat for all combinations of systems. (Cu-Zn, Cu-Ni, Cu-Pb, Zn-Ni, Zn-Pb, and Pb-Ni). If the potential is negative, switch the red and black plugs. Turn the multimeter off when done.

7. Complete Data Table 1 on the data sheet and answer the related questions.

Part 2

1. Remove the copper and zinc electrodes from the beakers and dry them thoroughly. Sand off any oxidation from the electrodes and wipe with a paper towel. Dip in 1 M HCl and then dry thoroughly.

2. Set up the cell as pictured in Figure 18.2.
- Turn a ring stand around and place a hot plate in front of the rod of the ring stand.
- Place a 1000 mL beaker on the hot plate and attach two test tube clamps to the ring stand, so that the bottom one is just above the lip of the beaker.
- Insert two 70 mL test tubes into the clamps so that the bottom of each test tube is about an inch from the bottom of the beaker.
- Suspend a thermometer into the beaker with a thermometer clamp, making sure it does not touch the bottom of the beaker.
- Fill the beaker with 900 mL of tap water (using a smaller beaker), and then fill the cells with the 0.1 M zinc nitrate and 0.1 M copper nitrate solutions (from the beakers in Part 1) so that the level of the solution in each test tube is lower than the level of the water in the 1000 mL beaker.

3. Remove a 12" piece of string (salt bridge) that has been immersed in 0.1 M KNO_3 or 0.1 M $NaNO_3$. Gently blot it on a paper towel to remove excess solution. Insert the salt bridge to connect the cells. The salt bridge should be immersed in the solutions at least a ½ inch. It is acceptable for the string to adhere to the sides of the test tubes.

4. Insert the electrodes (<u>with leads attached to the top of the electrodes</u>) into their respective solutions, and rest the multimeter on the clamps. Make sure the black plug is attached to the anode port and the red plug is attached to the cathode port. Use the electrodes to position the string, so that the string (salt bridge) is not touching either electrode. DO NOT TOUCH THE APPARATUS AFTER THIS POINT.

Figure 18.2. Set up for ΔH and ΔS determination

5. Set the multimeter to DC.   Start heating using a setting of 5, recording the temperature and voltage at approximately 2–5°C intervals.   The voltage should decrease as the temperature increases for this combination of cells.

6. When the temperature reaches 50°C, disconnect the leads at the multimeter. Remove and dry the electrodes.   Return the solutions in the test tubes to their respective beakers and allow the solutions to cool for Part 3. Turn the multimeter off.

7. Complete Data Table 2 and plot ΔG versus K. This can be done on the computer using the directions below. Determine ΔS and ΔH and answer the related questions.

Graph of ΔG vs. K to determine ΔH and ΔS
1. Go to Start, All Programs, Microsoft Office, Excel.
2. Enter "Voltage" in the first column and "Temp in C" in the second column.
3. Enter the data from Table 2.
4. Enter "T (K)" in the first line of the third column.
5. In the second line of the third column (cell C2), type the following formula: =B2+273. (B2 can be typed in or entered with the cursor.) Press Enter. Select the cell again (C2) and move the cursor over the small black square in the bottom right hand corner. The cursor should turn into a plus sign. Click and drag until the last row of data. Release the cursor.
6. Enter "G (kJ)" in the first line of the fourth column.
7. In the second line of the fourth column (cell D2), type the following formula: =−(2*96500*A2)/1000. (A2 can be typed in or entered with the cursor.) Press Enter. Select the cell again (D2) and move the cursor over the small black square in the bottom right hand corner. The cursor should turn into a plus sign. Click and drag until the last row of data. Release the cursor. Highlight the data, right click, and choose Format Cells. Click the Number tab and select Number from the Category list. Set Decimal Places to 1. Click OK.
8. Select the data in columns 3 and 4 (T in K and G) and click the Insert tab. Click the arrow under Scatter, choose the option "Scatter with smooth lines and markers". Move the chart underneath the data. Note the region that looks fairly linear with an <u>almost-flat</u> but positive sloping line. (The values should be about ± 200 units of those calculated in Error Analysis question 6.)
9. Highlight the range of points noted, and replot the data with these points (step 8). This may have to be repeated once more.
10. Right click a value on the y-axis, click Format Axis. Under "Horizontal axis crosses", check the box "Axis value" and input the lowest value on the y-axis (the biggest negative number). Click close.
11. Right click on one of the data points, click Add Trendline and select Linear. Check the boxes to "Display Equation on chart" and "Display R-squared value on chart". Close.
12. Under Chart Tools, click the Layout tab.
    a. Click Chart Title, click Above Chart. Add a title.
    b. Click the Axis Titles icon. Click Primary Horizontal Axis Title, click Title Below Axis. Add the x axis title.
    c. Click the Axis Titles icon. Click Primary Vertical Axis Title, click on Rotated Title. Add the y axis title.
13. Delete the Legend.
14. Right click on any horizontal line in the chart. Select Delete.
15. Move the box containing the equation of the line and R^2 value to any open area below the line.
16. Write the equation of the line and the R^2 value on Data Table 2.
17. If capabilities in the lab allow print one graph for each partner. If not, show the graph (on the computer screen) to the instructor.
18. Exit, and DO **NOT** SAVE.

Part 3
1. Thoroughly dry the copper and zinc electrodes. Sand off any oxidation from the electrodes and wipe with a paper towel. Dip in 1 M HCl and then dry thoroughly.

2. Fill a beaker ¾ full with 0.01 M copper nitrate.

3. Construct a system of 0.1 M zinc nitrate (from Part 1) and 0.01 M copper nitrate. Place the zinc and copper electrodes in their respective solutions. Make sure these are at room temperature.

4. Remove a 6" piece of string that has been immersed in 0.1 M KNO_3 or 0.1 M $NaNO_3$. Gently blot it on a paper towel to remove excess solution. Insert the salt bridge to connect the cells, so that it is not touching either electrode.

5. Attach the black plug to the anode and the red plug to the cathode.

6. Record the voltage.

7. Repeat steps 2 – 6 using 1.0 M copper nitrate in place of 0.01 M copper nitrate.

8. Remove and dry the electrodes.   Turn off the multimeter.

9. Complete Data Table 3 on the data sheet and answer the related questions.

LABORATORY REPORT SHEET

Name _______________________________   Date _______________________

Data Table 1. Measure voltages of different cells, calculate ΔG and K.   Temperature: _________
(All concentrations are 0.10 M)

	E (V)	ΔG ($\Delta G = -n\,F\,E$) Work does not need to be shown. (kJ/mol)	K ($\Delta G = -RT\,\ln K$) Work does not need to be shown.
Cu-Ni cell			
Reaction:			
Cell Diagram:			
Cu-Pb cell			
Reaction:			
Cell Diagram:			
Cu-Zn cell			
Reaction:			
Cell Diagram:			
Ni-Pb cell write and record as a voltaic cell			
Reaction:			
Cell Diagram:			
Ni-Zn cell			
Reaction:			
Cell Diagram:			
Pb-Zn cell			
Reaction:			
Cell Diagram:			

Data Table 2. Determination of ΔH and ΔS.

Balanced redox equation for the cells: _______________________________________

Cell diagram: ____

Temperature, °C	Temperature, K [*]	E, V	$\Delta G\ (= -n\,F\,E)$, kJ/mol [*]

* These columns do not need to be calculated if using the computer to graph the data.

Equation of line ____ R^2 ________________

ΔS _________________________________ ΔH ___________________________

Data Table 3. Measure the voltage of different concentrations of cells.

<u>0.01 M copper nitrate</u>

Reaction: ____

Cell Diagram: ____

E, measured experimentally: __________________

E, calculated: (Use Nernst Equation)
     For $E°$, use the E from Part 1

<u>1.0 M copper nitrate</u>

Reaction: ____

Cell Diagram: ____

E, measured experimentally: __________________

E, calculated: (Use Nernst Equation)
     For $E°$, use the E from Part 1

	ΔG, kJ/mol	ΔG, kJ/mol
	Measured Experimentally	Calculated from the Nernst Equation
0.01 M $Cu(NO_3)_2$		
0.10 M $Cu(NO_3)_2$		
1.0 M $Cu(NO_3)_2$		

Error Analysis Questions:

For questions 1-4, complete the sentences:

1.   As E becomes more positive, ΔG will ________________________________.

2.   As E becomes more positive, K will ________________________________.

3.   When using the Nernst equation for a voltaic cell, if the concentration in the anode cell is decreased, the value of E will become ______________ and the value of G will become more ______________.

4.   When using the Nernst equation for a voltaic cell, if the concentration in the anode cell is increased, the value of E will become ______________ and the value of G will become more ______________.

5.   Why are voltaic and electrolytic cell reactions compared using voltages instead of K values?

6.   Using the following published data, calculate the theoretical values of ΔH and ΔS of the Zn-Cu cell system. Stoichiometric multipliers and units must be used for credit.

	Zn	Zn^{2+}	Cu	Cu^{2+}
ΔH, kJ/mol	0	−152.4	0	51.88
ΔS, J/mol-K	41.6	−106.48	33.3	−26.4

ΔH:

ΔS:

Name________________________________

<u>HOMEWORK – There are FOUR pages to this homework</u>

1. The following electrodes and solutions are available for an experiment at 25°C:

Sn, Mn, KNO_3 (0.1 M), $Sn(NO_3)_2$ (1.0 M), and $Mn(NO_3)_2$ (1.0 M)

Complete the following for a <u>**VOLTAIC**</u> cell:

Sn: Half-cell reaction:	
Mn: Half-cell reaction:	
Anode __________ Cathode ______________	
Balanced net ionic equation	
Cell Diagram	
$E°$	
$\Delta G°$	
K	

This reaction is spontaneous because

$E°$ is _____, ΔG is _____, and K is _________. The equilibrium lies with the ______________.
      +/−   +/−   large/small   reactants/products

Calculate the new E_{cell} if 0.1 M $Mn(NO_3)_2$ was used (all other parameters remain the same).

Calculate the new E_{cell} if 2.0 M $Mn(NO_3)_2$ was used (all other parameters remain the same).

Generalizations for the Nernst Equation:
If the concentration of the products decrease in a voltaic cell, E __________________.

If the concentration of the products increase in a voltaic cell, E __________________.

2.   The following electrodes and solutions are available for an experiment at 25°C:

Al, Zn, KNO_3 (0.1 M), $Al(NO_3)_3$ (1.0 M), and $Zn(NO_3)_2$ (1.0 M)

Complete the following for an **ELECTROLYTIC** cell:

Al: Half-cell reaction:	
Zn: Half-cell reaction:	
Anode __________ Cathode ______________	
Balanced net ionic equation	
Cell Diagram	
$E°$	
$\Delta G°$	
K	

This reaction is *not* spontaneous because

$E°$ is _____ , ΔG is _____ , and K is ________ . The equilibrium lies with the ____________ .
    +/–   +/–   large/small   reactants/products

Calculate the new E_{cell} if 0.1 M $Al(NO_3)_3$ was used (all other parameters remain the same).

Calculate the new E_{cell} if 2.0 M $Al(NO_3)_3$ was used (all other parameters remain the same).

Generalizations for the Nernst Equation:
If the concentration of the reactants decrease in an electrolytic cell, E ____________________ .

If the concentration of the reactants increase, in an electrolytic cell, E ____________________ .

3.   An experiment containing the following components was performed where the entire <u>voltaic</u> cell system was heated:

$$Ag, Mg, KNO_3 \ (0.1 \ M), \ AgNO_3 \ (1.0 \ M), \ and \ Mg(NO_3)_2 \ (1.0 \ M)$$

Complete the following:

Ag: Half-cell reaction:	
Mg: Half-cell reaction:	
Anode __________ Cathode ______________	
Balanced net ionic equation	
Cell Diagram	
$E°$	
$\Delta G°$	
K	

Temperature, °C	Temperature, K	E, V	$\Delta G \ (= - n \ F \ E)$
27.0		3.16 V	
47.0		3.14 V	
67.0		3.12 V	
87.0		3.09 V	

Determine ΔS and ΔH graphically. You may use Excel to graph the data; please put the equation of the line on the graph. Include units in the values below.

Equation of line: ___________________________ ΔS ______________ ΔH ______________

Use thermodynamic data tables (found in the Appendix of a Chemistry text) to calculate the values of ΔS and ΔH.

$\Delta S°$:

$\Delta H°$:

4.   Answer the following questions with one of the following eight answers (answers may be used more than once).

+	−
large	small
spontaneous	non-spontaneous
products	reactants

a. In an experiment, if E is positive, ΔG must be ___________, and K must be ___________. The reaction is ___________________ and the equilibrium lies with the ______________.

b. In an experiment, if the E is negative, ΔG must be ___________, and K must be ___________. The reaction is ___________________ and the equilibrium lies with the ______________.

c. If a reaction is exothermic, ΔH is ___________. For a reaction to be unconditionally spontaneous, ΔS should be ___________. However, if ΔS is ___________, the reaction is spontaneous only at low temperatures.

d. If a reaction is endothermic, ΔH is ___________. For a reaction to be unconditionally non-spontaneous, ΔS should be ___________. However, if ΔS is ___________, the reaction is spontaneous only at high temperatures.

EXPERIMENT 19
ELECTROLYSIS AND FARADAY'S LAWS

An apparatus consisting of two aqueous electrolysis cells connected in series will be set up. In one cell, copper is electrolytically transported from the anode to the cathode. In the other cell, $H_{2\,(g)}$ and $O_{2\,(g)}$ are the products formed at the two electrodes. Using the results obtained from the copper cell, Faraday's laws and the ideal gas law will be used to calculate the volume of hydrogen theoretically generated in the water cell during the electrolysis. This will be compared to the experimental volume.

Key Chemical Reactions"

<u>Copper cell</u>:	Anode:	$Cu_{(s)} \rightarrow Cu^{2+}_{(aq)} + 2\,e^-$
	Cathode:	$Cu^{2+}_{(aq)} + 2\,e^- \rightarrow Cu_{(s)}$
<u>Water Cell</u>:	Anode:	$H_2O_{(l)} \rightarrow \frac{1}{2}\,O_{2(g)} + 2H^+_{(aq)} + 2\,e^-$
	Cathode:	$2\,H_2O_{(l)} + 2\,e^- \rightarrow H_{2(g)} + 2\,OH^-_{(aq)}$

Key Mathematical Equations:

$$\text{g Cu} \times \frac{1 \text{ mole Cu}}{63.55 \text{ g Cu}} \times \frac{2 \text{ moles e}^-}{1 \text{ mole Cu}} \times \frac{1 \text{ mole } H_2}{2 \text{ moles e}^-} = \text{moles } H_2$$

$$P_{H_2} = P_{atm} - P_{H_2O} \qquad\qquad V = \frac{n_{H_2}RT}{P_{H_2}}$$

$$\% \text{ error} = \frac{\text{Experimental - Theoretical}}{\text{Theoretical}} \times 100$$

Discussion

In the first half of the 19th century, Michael Faraday discovered that the quantity of an element that is liberated at an electrode is directly proportional to the amount of electricity that is passed through the solution. The basic unit used to express the quantity of electricity is the Coulomb (C), the unit of charge. The charge on an Avogadro's number of electrons (one mole of electrons) is equal to 96,485 Coulombs. This quantity of electricity (often rounded off to 96,500 coulombs) is equal to one Faraday (F). Also, there is a convenient conversion factor between coulombs and amp-seconds.

$$96,485 \text{ C/mole e}^- = 1 \text{ F}$$

$$1 \text{ C} = 1 \text{ amp-second (1 A-s)}$$

A and s are BOTH in the numerator; they are multiplied together.

There are two different types of electrolysis cells. An electrolytic cell is one which requires electricity to proceed and a galvanic (voltaic) cell is one which generates electricity. In this experiment, two electrolytic cells are connected in series, so that whatever charge flows through one cell also flows through the other cell. A schematic diagram of the apparatus that is used in this experiment is shown in Figure 19.1. In an electrolytic cell, oxidation of the reactant takes place at the anode and reduction of the reactant takes place at the cathode. The anode and cathode reactions in this experiment are as follows for each cell:

Copper cell: Anode: $Cu_{(s)} \rightarrow Cu^{2+}_{(aq)} + 2e^-$

               Cathode: $Cu^{2+}_{(aq)} + 2e^- \rightarrow Cu_{(s)}$

Water Cell: Anode: $H_2O_{(l)} \rightarrow \frac{1}{2} O_{2(g)} + 2H^+_{(aq)} + 2e^-$

               Cathode: $2 H_2O_{(l)} + 2e^- \rightarrow H_{2(g)} + 2 OH^-_{(aq)}$

The calculation of the electrical charge that has been passed through a circuit during a particular interval of time can be obtained from the change in mass at any one of the four electrodes. The increase in mass of the cathode or decrease in mass of the anode in the copper cell, or the volume of hydrogen collected at the cathode or oxygen at the anode in the water cell, should all theoretically yield the same answer.

In this experiment, the difference in mass of the copper anode will be used to calculate the moles of H_2 produced. Using the ideal gas equation, the theoretical volume of H_2 is calculated. This is then compared to the actual volume of H_2 collected.

Example
Consider a circuit as shown in Figure 19.1, containing two cells in series. The equations for the electrolytic cells are as follows:

$$Cu \rightarrow Cu^{2+} + 2e^- \qquad\qquad 2 H_2O + 2e^- \rightarrow H_2 + 2 OH^-$$

After current has been passed through the circuit for a period of time, the anode in the copper cell is found to have decreased in mass by 2.020 g. During this time, 810. mL of H_2 is collected over water at 747.0 mm Hg and 27.0 °C at the cathode in the water cell. Calculate the theoretical volume of H_2 liberated based on this data, the percent error of the data from theoretical and the number of Faradays of electricity transported.

1. Calculate the number of moles of H_2 generated.

$$2.020 \text{ g Cu} \times \frac{1 \text{ mole Cu}}{63.55 \text{ g Cu}} \times \frac{2 \text{ moles e}^-}{1 \text{ mole Cu}} \times \frac{1 \text{ mole H}_2}{2 \text{ moles e}^-} = 3.181 \times 10^{-2} \text{ moles H}_2$$

2. Calculate the partial pressure of H_2. The vapor pressure of water must be accounted for using Dalton's Law of Partial Pressures since the hydrogen is collected over water. The Appendix contains the vapor pressure of water at various temperatures.

$$P_{H_2} = P_{atm} - P_{H_2O} \qquad\qquad P_{H_2} = 747.0 \text{ mm Hg} - 26.7 \text{ mm Hg} = 720.3 \text{ mm Hg}$$

3. Using the ideal gas equation, rearrange and solve for the theoretical volume of H_2.

$$V = \frac{n_{H_2} RT}{P_{H_2}} = \frac{(3.181 \times 10^{-2} \text{ mol})(0.0821 \frac{\text{L-atm}}{\text{mol-K}})(300.2 \text{ K})}{0.9478 \text{ atm}} = 0.8272 \text{ L}$$

4. The actual yield of H_2 is 810. mL. The percent error is

$$\% \text{ error} = \frac{\text{Actual} - \text{Theoretical}}{\text{Theoretical}} \times 100 \qquad = \frac{0.810 - 0.8272}{0.8272} \times 100 = -2.08\,\%$$

The minus sign shows that the actual amount of H_2 collected is less than the theoretical.

5. Calculate the number of Faradays of electricity transported. Refer to the reactions at the anode and cathode to obtain the correct number of electrons transferred per mole of copper.

$$2.020 \text{ g Cu} \times \frac{1 \text{ mole Cu}}{63.55 \text{ g Cu}} \times \frac{2 \text{ moles e}^-}{1 \text{ mole Cu}} \times \frac{1 \text{ Faraday}}{1 \text{ mole e}^-} = 6.362 \times 10^{-2} \text{ F}$$

The coulombs of electricity can be calculated from this number. The current (in amps) passing through the cells can also be calculated if the time of electrolysis is known.

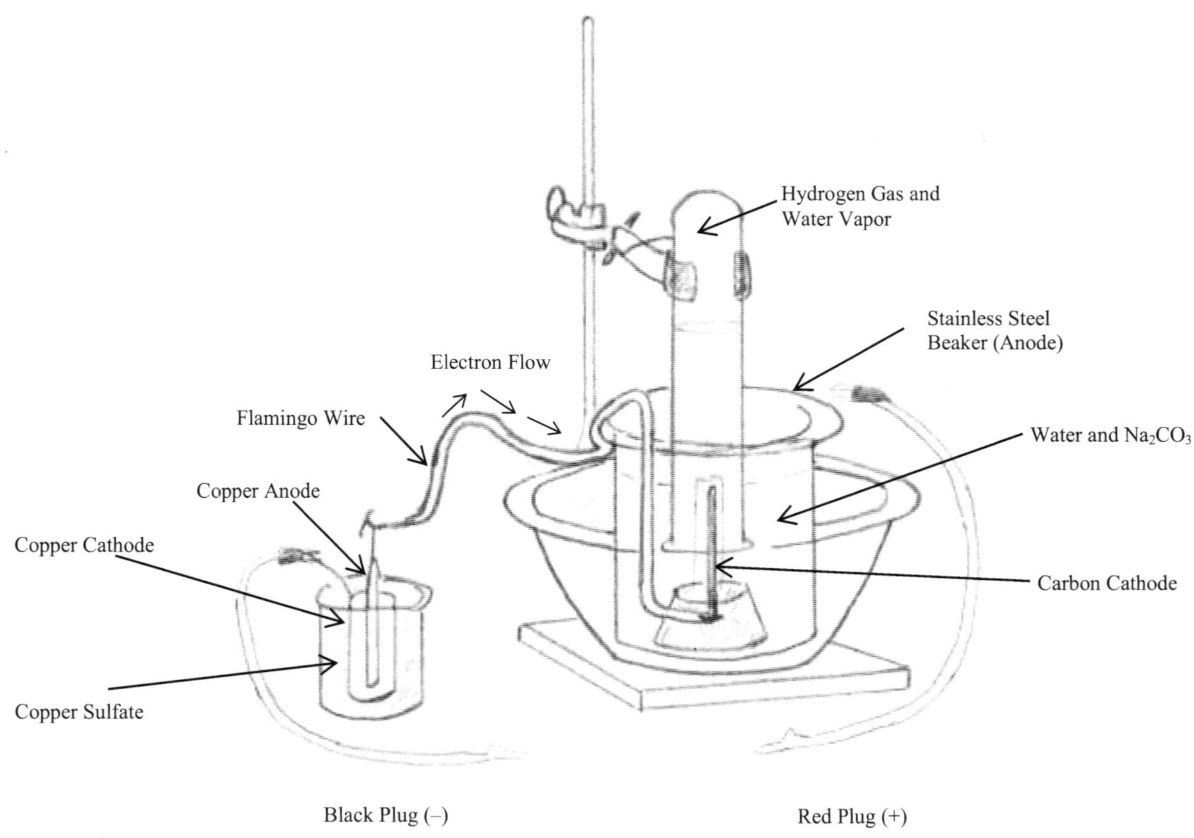

Figure 19.1
Electrolysis Set Up

Chemicals

Copper	$Cu_{(s)}$
Copper (II) sulfate	$CuSO_{4(aq)}$
Sodium carbonate	$Na_2CO_{3(s)}$
Hydrogen (formed)	$H_{2\ (g)}$

Waste

Water waste can be disposed down the drain. Used copper (II) sulfate solution can be recycled; pour any solution back into the copper (II) sulfate bottle through a filter.

Procedure

Steps 1-7 detail how to set up the apparatus as shown in Figure 19.1. Also attach an aspirator to remove water from the bowl.

1. Fill a dry 100 mL beaker about ½ full of Na_2CO_3. Transfer this into the stainless steel beaker and fill to about ¼ inch from the top with tap water. Stir until most of the salt is dissolved. Place the stainless steel beaker in the plastic bowl.

2. Assemble the carbon electrode apparatus. Sand the exposed copper of the formed coated wire ("flamingo wire") and then insert the right angle end of the flamingo wire into the side hole of the rubber stopper. Exposed copper wire should be visible from the top hole of the stopper. Next, firmly press the <u>flat</u> end of the carbon electrode into the top of the rubber stopper (if one end is not flat, contact the instructor to sand it). Make sure the exposed wire and carbon electrode have good contact with each other and the electrode is very secure in the stopper. Make sure the flamingo wire is able to stay in place and does not fall to one side. Place the wire mesh screen over the carbon electrode. It is not necessary to place it in the groove on the stopper.

3. Place the carbon electrode assembly in the stainless steel beaker, making sure the flamingo wire is securely hooked over the rim of the beaker. Add tap or DI water to the beaker so that the water level is about ¼ inch from the rim.

DO NOT CHANGE THE SHAPE OF THE FLAMINGO WIRE! IT IS FORMED SO
THAT NO WATER CAN TRAVEL TO THE COPPER CELL.

4. Fill the gas collection tube with tap water, and then use the DI water bottle to completely fill the tube, making a dome of water. Place the fat end of a large rubber stopper over the top of the tube and invert the tube into the steel beaker. When the lip of the collection tube is just below the surface of the water, remove the rubber stopper. Lower the collection tube 2 – 3 inches over the carbon electrode assembly and clamp the tube in place. There should be no or a very little air bubble in the tube.

5. Weigh the copper anode and record its mass on the data sheet. The cathode is too heavy to be weighed.

6. Assemble the copper-plating cell so that the anode is in the middle of the cylindrical cathode and is hanging from the loop of the flamingo wire. It should not be touching

the bottom of the beaker. Pour 1 M copper (II) sulfate solution into the 150 mL beaker until the cathode is just covered. It may be necessary to elevate the beaker on a stopper so that a good portion of the electrode is covered with copper (II) sulfate. Make sure that the two copper electrodes are not in contact with each other. If they are, they will show no change in mass although H_2 will keep coming off in the water cell.

7. Ensure the power switch to the main power supply unit is shut off and then plug in the power supply unit. Plug in both the black and red leads into the power supply. Clip the black (negative) alligator clip to the copper cathode, making sure the copper cell is stabilized so that the copper anode and cathode are not touching. Do not clip the red alligator clip to the beaker at this point. Call the instructor before proceeding with the next step to have them check the set up.

8. When the instructor has approved the set-up, clip the red (positive) alligator clip to the stainless steel beaker. The apparatus is now ready for electrolysis.

DO NOT TOUCH THE WATER OR ANYTHING METAL ONCE THE EXPERIMENT IS COMPLETELY CONNECTED.

9. Turn the power on to the power supply. Adjust the voltage and amperage to obtain a current flow of 0.7 to 0.8 A. If the current is too high, the electrolysis will take place more rapidly but poor results may be obtained. Collect hydrogen until the aqueous level in the gas collection tube is at the level of the water in the beaker. Aspirate water as necessary from the bowl. Be careful not to displace more than the tube can hold. Break the circuit by disconnecting either of the terminal alligator clips.

10. Very carefully remove the copper anode and rinse it with deionized water and then, if instructed, with acetone. When it is completely dry, weigh it and record the mass. The gain in mass of the cathode should equal the loss in mass of the anode; however, the cathode is too heavy for the balances.

11. Return the $CuSO_4$ solution through a filter to the laboratory stock bottle.

12. Carefully adjust the height of the tube so that the liquid level inside the tube is the same as the liquid level in the steel beaker. Mark the tube at the liquid-gas junction. Remove the gas tube from the beaker, drain all of the liquid from it, and determine the volume of wet hydrogen collected to within ± 1 mL by utilizing the mark on the tube and a 1000 mL graduated cylinder.

13. Take the temperature of the water, and read the barometer to obtain the pressure in the lab.

14. Discard the Na_2CO_3 solution as instructed by your instructor. This solution is extremely messy and will coat the carbon electrode if the electrode is not thoroughly washed. Everything that was in contact with the Na_2CO_3 must be washed and scrubbed thoroughly.

290

LABORATORY REPORT SHEET

RECORD ALL VALUES WITH PROPER UNITS!

Name____   Date___________

Mass of copper anode before electrolysis ______________   Time start ________________

Mass of copper anode after electrolysis ______________   Time end ________________

Mass of copper removed ______________   Total time ______________

Use units on all values!	Measured Value	Value with units that match R $0.0821 \ \dfrac{L \cdot atm}{mole \cdot K}$
Temperature		
Atmospheric pressure		
Vapor pressure of water *Show interpolation calculation*		
Pressure of H_2		

Moles of H_2 generated, based on the mass of copper removed:

Theoretical volume of H_2 (show all work):

Experimental volume of H_2 collected: ______________

% Error between the theoretical and experimental volumes of H_2:

Continued on back:

Calculate the following based on the mass of copper removed (or moles of H_2 generated):

Faradays of electricity transported:

Coulombs of electricity transported:

Current (in amps) passing through the cells:

Approximate average value of amperage: ____________________ A
(from readout on the power supply)

Error Analysis

1. Using the experimental volume of hydrogen collected, calculate the grams of copper that would need to be removed to correspond to this volume.

2. The experimental and theoretical volumes should be the same. In complete sentences, explain two additional sources of error, stating if the error would affect the experimental or theoretical volume of hydrogen, and exactly how it would affect the experimental or theoretical volume of hydrogen.

Example: If the _amperage is too high_, hydrogen will be generated too quickly, and _too much hydrogen will be generated_, displacing an amount greater than the water in the bottle, a total volume that is _unable to be collected_. The _experimental volume will be affected and it will appear smaller than the theoretical volume_.

Error #1:

Error #2:

Name _______________________________

HOMEWORK EXERCISES

1. A current of 0.50 amperes flowed through a cell for two hours in which $CuSO_4$ is the electrolyte and a copper electrode is present.

    a. Write the oxidation reaction (anode) for the cell.

    b. Write the reduction reaction (cathode) for the cell.

    c. How many Coulombs of electricity are generated?

    d.   How many moles of electrons are generated?

    e. How many Faradays are generated?

    f. Calculate the change in mass of the copper electrode after two hours.

2. a. What is the electrical current (in amps) passing through a $CuSO_4$–Cu cell if it takes 2 hours to transfer 0.7550 grams of copper from the anode to the cathode?

Continued

b. Calculate, by interpolation, the vapor pressure of water at 23.4 °C. Refer to the Appendix for the vapor pressure of water at 23 and 24 °C.

c. Calculate the theoretical volume of H_2 produced simultaneously at the cathode of the H_2O-H_2 cell during this time. The H_2 is collected over water at 752.6 mm Hg and 23.4 °C.

d. Calculate the % error if 280. mL of H_2 are collected.

EXPERIMENT 20
CSI: SMU

This is a FUN lab highlighting organic compounds in a forensic setting.
- There is no quiz or prelab due for this experiment.
- The take-home assignment is due at the beginning of your lab period the day this experiment is performed. The take-home assignment is worth 80 points. When the procedures are completed, hand in the completed data sheet, worth 20 points.

Synthesis of Oil of Wintergreen

1. Fill a 150 mL beaker over half full with water. Place on a hot plate and heat to boiling. Replenish water as necessary during this experiment to keep the water level above half full. Continue with steps 2 and 3 during heating.

2. Tare a small test tube (13 x 100 mm) with a 100 mL beaker. Add approximately 0.2 g of salicylic acid to the test tube. Add 25 drops of methanol and 10 drops of 6 M H_2SO_4. Mix with a stirring rod. The alcohol is the excess reagent and the carboxylic acid is the limiting reagent.

3. Add 4–5 rounded scoops of $NaHCO_3$ (about 4 g) to a 100 mL beaker and add about 50 mL of water. Mix until the $NaHCO_3$ is almost completely dissolved.

4. When the water is boiling, place the test tube in the boiling water and let the mixture heat for 7 – 10 minutes. The mixture will most likely be solid at this time.

5. Using test tube tongs, remove the test tube from the boiling water and add DI water so the test tube is half full. Mix. Pour the contents of the test tube into the $NaHCO_3$ solution. Rinse the test tube with DI water and add to the $NaHCO_3$ solution. Mix. After all the bubbles have subsided, gently waft the vapors from the beaker to smell.

Synthesis of Banana Oil

1. Replenish the water in the 150 mL beaker as necessary during this experiment to keep the water level above half full. Continue with steps 2 and 3 during heating.

2. Add 25 drops of isopentyl alcohol and 10 drops of 6 M acetic acid in a small test tube. Add 10 drops of 6 M H_2SO_4. Mix. The alcohol is the excess reagent and the carboxylic acid is the limiting reagent.

3. Add 4–5 rounded scoops of $NaHCO_3$ (about 4 g) to a 100 mL beaker and add about 50 mL of water. Mix until the $NaHCO_3$ is almost completely dissolved.

4. When the water is boiling, place the test tube in the boiling water and let the mixture heat for 7 – 10 minutes.

5. Using test tube tongs, remove the test tube from the boiling water and pour the contents of the test tube into the $NaHCO_3$ solution. Rinse the test tube out with DI water and add this to the $NaHCO_3$ solution. Mix. After all the bubbles have subsided, gently waft the vapors from the beaker to smell.

<u>Latent Fingerprint Analysis</u>

<u>Ninhydrin</u>
1. Lightly coat (using a small paint brush) the entire surface of the paper with ninhydrin solution and then let the paper dry for at least 10 minutes.

2. Place the paper, print side up, between paper towels (at least two thicknesses) and iron (heat set   to low steam) for 2 seconds. Repeat in 2 second intervals until complete prints form.

3. Compare print to known fingerprints.

<u>Cyanoacrylate (Superglue) Fuming</u>
Handle the can carefully – by the rim only!
1. You will share a fuming chamber (empty paint can) with another lab bench, so that two cans are fumed in one chamber. Place the soft drink cans in the empty fuming chamber (leave one tab up, one down to differentiate). Add 20–25 drops of superglue to the top of each drink can. Place a wet paper towel in the paint can. Replace the fuming chamber cover, sealing tightly, so that no fumes can escape.

2. Heat above 80 °C in the oven for 1.5 hours.

3. Remove from the oven and cool. When cool, open carefully, AWAY from your face. Remove the drink can and compare print to known fingerprints.

<u>Blood Presence Analysis</u>

<u>Kastle–Meyer Test</u>   *(There is enough for everyone to do this test.)*
1. Moisten a swab with two drops of DI water.

2. Add 1–2 drops of 70% ethanol to the swab and then gently roll it in the presumed blood sample.

3. Add two drops of the Kastle–Meyer solution. (If the solution turns pink, the solution has been oxidized and the test will be invalid.)

4. Add two drops of the 3% hydrogen peroxide solution. If a pink color develops <u>immediately</u>, then the presence of blood is confirmed. (Note: the color will change after 30 seconds even if blood is not present – this is due to the hydrogen peroxide oxidizing the phenolphthalein in the Kastle–Meyer solution.)

<u>Luminol/Luminol Analog Test</u>
1. Spray luminol solution over the suspected area.

2. Turn off the lights and observe for 1–2 minutes. Observe and record.

Take-Home Assignment   Name _________________________________

This assignment is five pages. Please staple them together.

Part 1. <u>Organic Functional Groups and Naming</u>. Complete the following table:

Functional Group Name	Functional Group Structure	Example	Name
Alcohol	$-OH$	CH_3OH	Methanol
Aldehyde			
Ketone			
Carboxylic acid	$$\begin{array}{c} O \\ \parallel \\ R-C-O-H \end{array}$$		
Ester			
Ether			
Amine	1°		
	2°		
	3°		
Amide			

Part 2. Synthesis of an Ester.

Esters are compounds of the form RCOOR′. They occur naturally in many plants and animals, and can be synthesized. They have a pleasant aroma usually described as "fruity" or "flowery".

In a Fisher esterification, an ester is formed from the reaction of a carboxylic acid and an alcohol with a strong acid as a catalyst. The OH contained in the carboxylic acid group is replaced with the alkoxy group (O–R′) from the alcohol, and an ester and water are formed.

$$R-\overset{\overset{\displaystyle O}{\|}}{C}-O-H \; + \; H-O-R' \quad \underset{}{\overset{H^+}{\rightleftharpoons}} \quad R-\overset{\overset{\displaystyle O}{\|}}{C}-O-R' \; + \; H_2O$$

Questions *Oil of Wintergreen*:

1. Draw the chemical reaction between salicylic acid (the limiting reagent) and methanol.
    - Box the major functional group of each organic compound.
    - The formula of the catalyst is in the box above the reaction arrow.

$\boxed{H_2SO_4}$

Salicylic acid + Methanol $\underset{catalyst}{\rightleftharpoons}$ Methyl salicylate + Water

Functional Group: ____________   ____________   ____________

2. What are two uses of methyl salicylate?

3. Comment on the toxicity of methyl salicylate. What is considered a lethal dose, and what is its uptake avenue (inhalation, skin contact, ingestion, etc.)?

4. Why is the mixture reacted with a sodium bicarbonate solution? Write the reaction. It does NOT react with the limiting reagent since limiting reagents are completely consumed.

Questions *Banana Oil*:

1. Draw the chemical reaction between acetic acid (the limiting reagent) and isopentyl alcohol.
    *   Box the major functional group of each organic compound.
    *   The formula of the catalyst is in the box above the reaction arrow.

$$\boxed{HCl}$$

Acetic acid   +   Isopentyl alcohol $\rightleftharpoons$ Isopentyl acetate   +   Water

catalyst

Functional Group: ________________   ________________   ________________

2. What is the IUPAC name for isopentyl alcohol? ________________________________

3. Banana oil is sold in ampoules. For what purpose are these ampoules used?

4. In this experiment, glacial acetic acid is used. What does "glacial" mean in this context?

5. What is the IUPAC name for acetic acid? ________________________________

6. Methyl anthranilate is an ester that is used as grape flavoring. It can be synthesized using a Fisher esterification. Draw the chemical reaction between the carboxylic acid and the alcohol. Write the formula of the catalyst above the reaction arrow.

Name: ________________   +   ________________ $\rightleftharpoons$ Methyl anthranilate   +   Water

Functional Group: carboxylic acid   alcohol   ester

The ester contains another organic functional group. What is it? ________________________

Part 3. Detection of Fingerprints.

Latent Fingerprint Questions

1. What is the definition of a latent fingerprint?

2. What does AFIS stand for? ____

3. There are two tests to expose latent fingerprints. One is ninhydrin and the other is cyanoacrylate. Ninhydrin is used on ________________ surfaces, and cyanoacrylate is used on ________________ surfaces. In the tests, the ninhydrin and the cyanoacrylate react with the ______________________________ from the skin in the fingerprint.

4. Ninhydrin was used to develop fingerprints in the year ____________ and cyanoacrylate was used to develop fingerprints in the year ______________. Cyanoacrylate has an interesting history. What was it used for on the battlefield?

5. Cyanoacrylate forms a polymer with the fingerprints. Draw the structure of cyanoacrylate and also draw the polymer that forms.

Structure of Cyanoacrylate

Structure of Polymer

Part 4. Detection of Blood.

Presumptive Blood Tests

1. What does presumptive mean in this context?

2. There are two tests that can be performed. One is Kastle-Meyer test and the other is the luminol test. For what situation should each be used?

3. The Kastle-Meyer test was developed in the year _________________________. It is extremely

sensitive, and can detect blood in concentrations as low as _________________________.

However, there are some things other than blood that can show a positive test, one of which

is _________________________.

4. Clearly explain how the Kastle-Meyer test works chemically. Be sure to include correct chemistry terms when describing the functions of the blood and hydrogen peroxide.

5. The luminol test was developed in the year _________________________. It is also an

extremely sensitive test, and can detect blood in concentrations as low as _______________.

However, there are some compounds other than blood that can show a positive test, one of

which is _________________________.

6. Clearly explain how the luminol test works chemically. Be sure to include correct chemistry terms when describing the functions of the blood and hydrogen peroxide, and a clear description how light is generated.

APPENDIX 1: HOW TO USE THE BALANCE, BUNSEN BURNER, BURET, AND PIPET

A. <u>The Balance:</u>

Mass and weight are two different quantities. Mass is a fundamental property of an object and is independent of position or location; it measures the amount of matter present. Weight is the force with which that mass is attracted by the mass of the earth. The weight of an object is directly proportional to its mass, with the proportionality constant being the acceleration of gravity. If two objects are in essentially the same location, then when their weights are the same, their masses will also be the same. Therefore, the masses of objects are compared by "weighing" the objects.

The instrument used to measure mass in this laboratory is an electronic analytic balance that indicates the mass to the nearest 0.0001 g. Figure A1.1 shows a diagram of this balance.

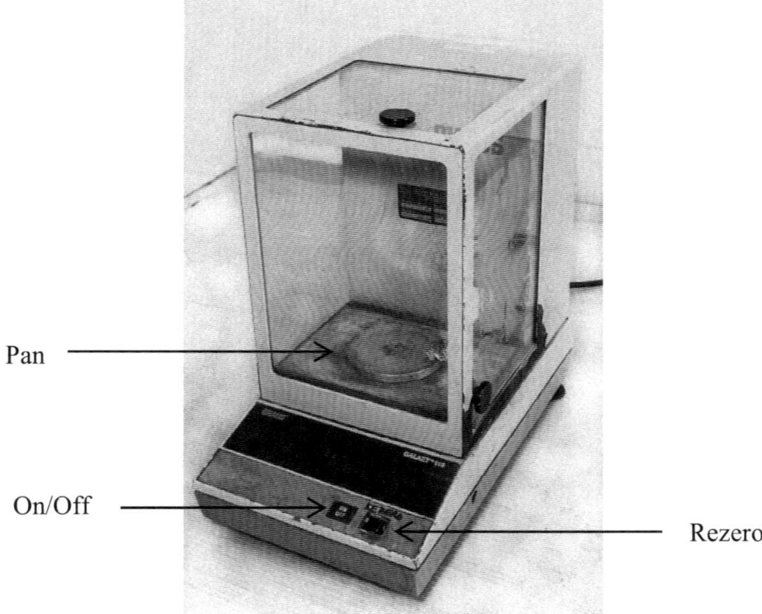

Figure A1.1

To operate the balance, check to see that the balance pan has nothing on it (no water or crystals), make sure that the balance doors are closed, and then press the ON/OFF switch (see Figure A1.1) if the balance is off. The display will show an array of 8s for about three seconds while the balance goes through a self-diagnostic check. If all functions are operating properly the balance will then display zero. If the balance has been in use, it will be on and the display should be reading zero with nothing on the balance pans. If it does not, press the REZERO switch to obtain a reading of zero.

Open either the right hand or left hand door of the balance and place the object to be weighed on the balance pan. Close the balance door, wait a few seconds for the balance to stabilize, and then record the mass which is displayed. Any object that is being weighed should <u>never</u> be placed directly on the balance pan unless the object is clean and dry. Powdered solids are to be weighed in a test tube or on a piece of weighing paper, and liquids are weighed in a flask or supported test tube.

There are two ways of determining the mass of a sample that requires a container or weighing square. In one method, which is often called "weighing by difference", the mass of the container or weighing paper is determined before and then after the sample is added. The mass of the sample is the difference between the two weighings.

The other method involves "taring", that is setting the balance to read zero by pressing the REZERO switch, with the empty container or weighing paper on the pan. The sample is then added to give a direct reading of the mass of the sample. This method is much more convenient than weighing by difference but it can be used only when the balance zero is not altered after taring until the final weighing is completed.

Important Points:

1. Keep the balance clean and uncontaminated at all times. Use stoppered containers when weighing volatile or corrosive substances.

2. When weighing powder samples, add very small increments of the sample from the tip of a spatula to the weighing paper until the desired mass is achieved. Do not begin by placing more than enough sample on the paper and then trying to remove the excess by scooping it up with a spatula. This can result in spillage or damage to the balance.

DO NOT RETURN MATERIAL FROM A CONTAINER BACK TO THE CONTAINER AFTER IT HAS BEEN IN THE PAN.

3. If sample is spilled, either on to the balance pan or the floor of the balance, it is your responsibility to clean it up immediately. If necessary, enlist the help of the TA. If cleaning involves the removal of the, the balance may be unstable for ten or fifteen minutes.

4. The maximum capacity of the balance is 110 g. If the mass on the pan is larger than 110 g the letter "H" will be displayed.

B. The Bunsen Burner:

The Bunsen burner (Figure A1.2) is the most commonly used source of heat in the laboratory. Natural gas or methane (CH_4) under a very low pressure from the city gas lines enters through the gas inlet (1). Under the burner is an adjusting screw (2) used to regulate the flow of gas. The gas enters the mixing chamber through a needle orifice at the base of the barrel. Air enters through openings (3) into the mixing chamber. An increased supply of air will result in a hotter flame. The openings may be regulated to vary the flame height (use the adjusting screw) and temperature (use the air regulator by twisting the barrel). The air–gas mixture emerges and burns at the top of the tube, converting the gas into carbon dioxide and water vapor. An increased supply of gas gives a larger flame.

To light the burner, close both needle valve and air vents completely by turning the gas adjusting screw (2) counterclockwise and the air inlet (3) clockwise as far as either will go. Open both the gas adjusting screw (2) and the air inlet (3) one full turn. Turn the main gas cock on full, parallel with the delivery nozzle. Ignite the emerging gas with the lighter. A small yellow flame will result which can now be adjusted with the aid of the two operating controls (never the main valve) to obtain a flame of the desired size and temperature. Adjust the flame size with the gas adjusting screw (2). A yellow flame with little or no inside cone indicates incomplete combustion of the gas and calls for an increase in the supply of air. Adjust the flame color to blue-violet with the air inlet (3).

With a properly adjusted burner there will be two distinct regions in the flame: a small blue cone just above the top of the barrel, and a much larger violet cone surrounding this. The hottest region in the flame is the blue inner cone.

Beakers, test tubes and crucibles being heated should be placed on a wire gauze support, slightly above the hottest portion of the flame – which is then spread around the object as efficiently as possible. Always adjust equipment before turning on the Bunsen burner.

When the burner is not in use always turn it off by closing the main gas cock. At the end of the laboratory period always check to make certain that the gas supply has been closed in this manner.

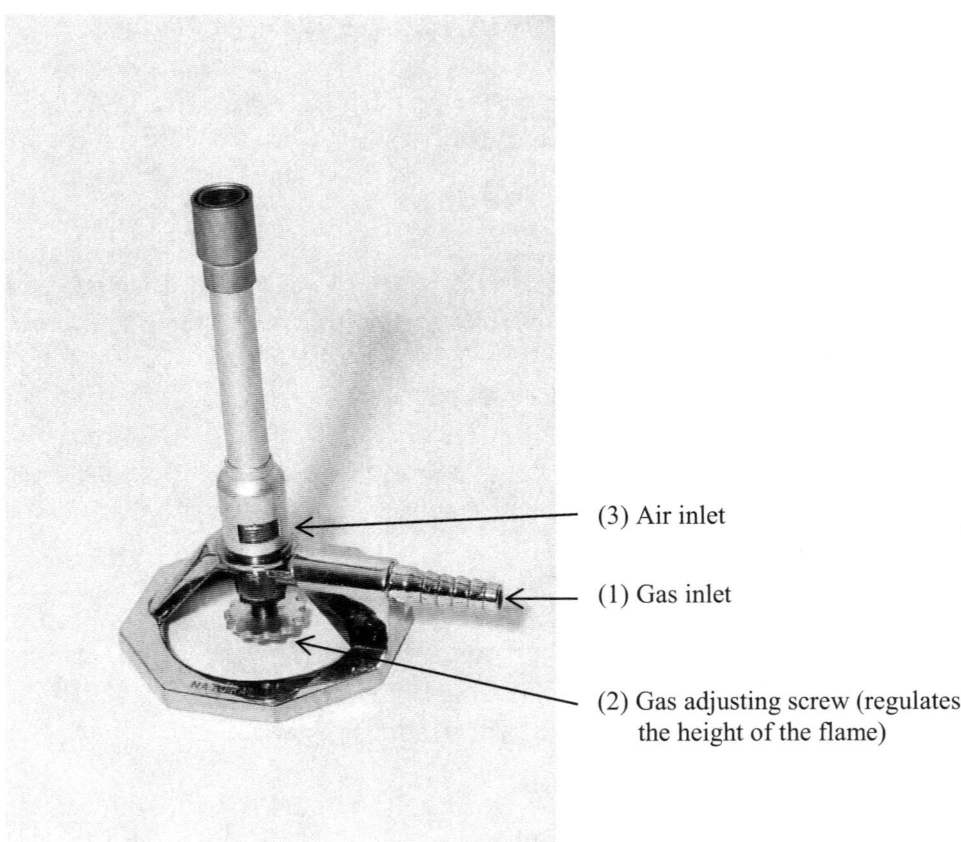

Figure A1.2

C. The Buret and Pipet

The buret (also spelled burette) is used to deliver a precisely known volume of a liquid. The most commonly used burets have volumes of 10 mL or 50 mL. Figure A1.3 shows pictures of a buret.

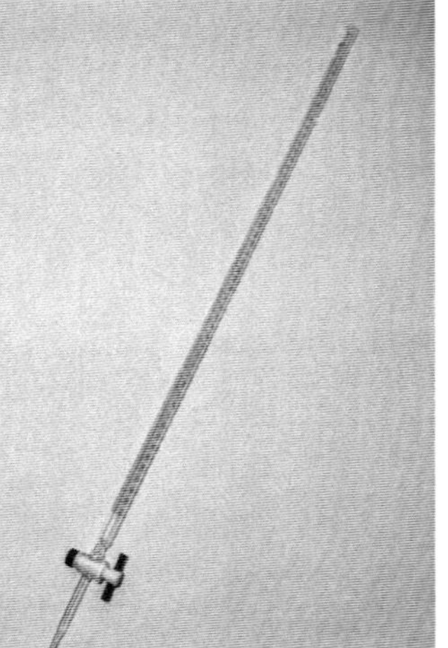

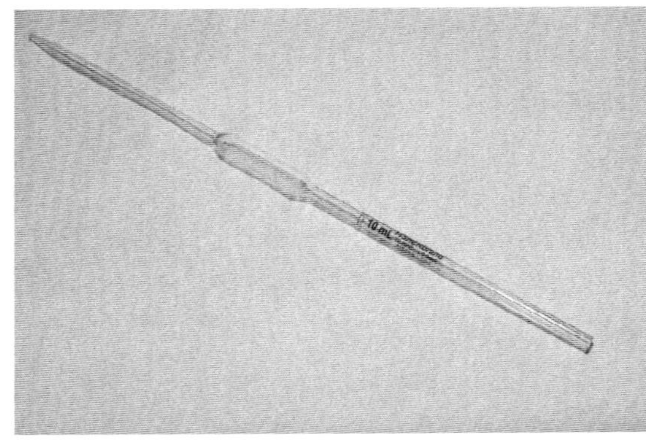

Figure A1.3   Figure A1.4

If the buret is dirty it will not drain properly. Before filling the buret with the liquid to be used, rinse the buret with a <u>small</u> portion of the solution to be delivered, wetting all areas of the inside wall and then allowing the liquid to drain through the stopcock to waste. The purpose of this rinsing, called normalization, is to make sure that the liquid which the buret will be finally filled with is not contaminated or diluted by any residue remaining in the buret from its previous use. Then, close the stopcock and, from a small beaker, fill the buret so that the level of the liquid is above the zero mL mark. Open the stopcock and allow the liquid to drain into a waste beaker until the liquid level is somewhere between the 0 and 1 mL mark. Make sure that there are no air bubbles in the stopcock area.

In a 50 mL buret, the scale on the barrel of the buret is marked off in one mL increments, from 0 to 50, with each mL further subdivided into 0.1 mL divisions. Read the position of the bottom of the curved surface of the liquid, the meniscus, to the nearest 1/10th, and then estimate to the hundredth or 0.01 mL. To make this reading it is helpful to hold a buret reader or a piece of paper with a dark mark on it in back of the buret close to the liquid level. This will cause the bottom of the meniscus to be outlined and make the position easier to determine. Be sure to place your eyes at the same level as the meniscus to minimize parallax.

Record this as the "initial buret reading" (remember, to the nearest 0.01 mL). Open the stopcock to dispense liquid. When done adding the liquid, wait a few seconds and then read and record, again estimating to the nearest 0.01 mL, the "final buret reading". The difference between the two buret readings is the volume of liquid delivered.

When finished using the buret, rinse it thoroughly with DI water and then completely fill it with DI water, draining the water through the stopcock. When rinsed, flip the buret, place it back in the buret clamp, and open the stopcock. Many of the solutions used in the buret will etch glass slowly and so burets which are not carefully rinsed can be ruined.

Like the buret, the volumetric pipet (see Figure A1.4) is used to deliver a precisely known volume of liquid. However, the volume delivered by the type of pipet used is fixed rather than adjustable. Pipets are filled by suction using a pipet pump or autopipetter.

To start, make sure that the pipet is clean and drains properly. Normalize the pipet by drawing up the solution to be delivered and then discarding it to waste.

<u>Pipet Pump (autopipetter)</u>: Gently insert a pipet into the pipet pump (autopipetter). Turn the wheel of the pipet pump with your thumb to draw up liquid until the bottom of meniscus is at the mark on the neck of the pipet. Depress the white bar on the side of the pipet pump to dispense. Touch the pipet tip to the beaker when done.

*** *Never let liquid enter the pipet pump. Always keep the level of the pipet pump <u>above</u> the level of the pipet.* ***

Since pipets are rated to deliver a specific amount of liquid, a small amount of the liquid will remain in the tip of the pipet after the desired volume has been discharged. Carefully rinse the pipet with deionized water at the end of the laboratory period.

APPENDIX 2: TECHNIQUES AND DEFINITIONS

<u>Aliquot</u>: A portion of a whole. A 10 mL portion of a 100 mL sample is a 10 mL aliquot.

<u>Centrifuging</u>: A piece of equipment that separates solids from the liquid. A small test tube centrifuge is used in the laboratory. Place a test tube of water or another sample directly across the sample to be centrifuged to ensure the centrifuge is balanced. Both test tubes should contain approximately the same volume, and should be filled no more than ½ inch from the top.

<u>Decanting</u>: In a solid-liquid mixture in which the solid has settled to the bottom of the vessel, the liquid can be decanted, or poured off, separating the liquid from the solid. To do this, transfer the liquid (or supernatant) slowly to avoid disturbing the solid. Some liquid may remain with the solid.

<u>Filtration</u>: In a cone funnel filter, fold the circular filter paper twice and open it up so that three thicknesses are on one side and one thickness is on the other. Place the paper in the filter and wet it with deionized water or as instructed. Slowly transfer the solution to be filtered, but never fill the funnel more than two thirds full. In a Büchner funnel, place the filter paper in the funnel, establish suction, and then wet the filter paper with the solution. Pour the solution in the middle so that solid is not collected on the filter walls.

<u>Graphing Data</u>: A well-constructed graph is very important in determining accurate data.
1. Select the two variables to be plotted, identifying which one is the dependent variable and which one is the independent variable. The dependent variable is the one that responds when a deliberate change is made in the other variable. For example, if NaOH is delivered to a solution in a titration and the pH is measured, the amount of NaOH delivered is the independent variable (it is deliberately changed), and the pH is the dependent variable, the response to that change.
2. Label the x and y axes, with the independent variable on the x axis and the dependent variable on the y axis. Include units.
3. Properly scale the axes. The graph area should take up most of the grid. To do this for each axis, take the range of the data and divide by the number of squares available. Choose major divisions that make smaller divisions easy to determine and data easy to interpret. The origin of each axis does not have to start at zero. For example, 210 mL of NaOH were delivered in a titration experiment, and this is the independent variable. The graph paper has 10 major divisions on the x axis with 10 smaller divisions within each major one. To scale, divide 210 by 10 (major divisions) to get 21. Round this UP to a convenient number, such as 25. Each major division is now 25 mL, and each minor division is 2.5 mL.
4. Draw a best fit curve for the data. If the data is supposed to be linear, draw a line that passes near most of the points. (It does not have to go through the centers of the points, and some points may be invalid.) If the data is supposed to be curved, use a French curve to draw a smooth line through the points.
5. Title the graph appropriately using the x and y axis labels. Include your name and date. For example, "Titration Curve of 0.15 M Acetic Acid with 0.25 M NaOH". The title should be in the form "y vs x", i.e., dependent value vs. independent value. Always include units.

<u>Meniscus</u>: In a buret or pipet, the liquid surface on the top will be convex. To properly read the volume, the level must be read at the bottom of the meniscus. Place your eye level with the bottom of the meniscus. It is often helpful to place a white piece of paper behind the buret or pipet to accurately read the volume.

<u>Normalize</u>: To rinse all surfaces of a vessel (buret, pipet, beaker) with the solution that it will contain. To normalize a buret, close the stopcock, add a small amount (5 – 10 mL) of the solution to the buret, and wet all areas of the inside wall by moving the buret. Drain the buret's contents by opening the stopcock and dispose of it properly. To normalize a pipet, draw up the solution into the pipet, and dispose of it in a waste container.

<u>Odor Testing</u>: Fan vapor with your hand towards your nose ("waft"). NEVER place your face over a vessel and ALWAYS check with your instructor prior to odor testing as some vapors can be toxic.

<u>pH Testing</u>: Insert a clean stirring rod into the solution and then touch the stirring rod to a small piece of pH paper. Compare the color on the paper to the guide provided on the pH paper container to evaluate the pH. <u>DO NOT PLACE PH PAPER INTO A SOLUTION TO TEST THE PH</u>.

<u>Precipitate</u>: A solid that has formed in a reaction.

<u>Titration</u>: Normalize a buret with the solution that will be delivered. Drain this solution into an Erlenmeyer flask containing the solution to be titrated. During delivery, gently swirl the flask to mix the two solutions. It is often helpful to put a piece of white paper under the flask to clearly observe the color of the solution in the flask. Use one hand on the stopcock to control delivery and the other to swirl the flask. Deliver the solution from the buret quickly at first, then very slowly near the end point, one or two drops at a time until a faint color persists for 15 or more seconds. Touch the buret tip to the inside wall of the Erlenmeyer flask, rinse the walls down with a small amount of DI water, and then read the volume on the buret.

<u>Weighing</u>: To weigh a sample of a solid, fold a square piece of weighing paper in fourths – corner to corner, and then repeated with the remaining two corners. Unfold and add the sample to the center to weigh. Remove the paper from the balance carefully using one of the upturned corners. NEVER put solid or liquid taken from a stock container back into the container.

APPENDIX 3: THE CHEMISTRY LABORATORY REPORT

A well-organized, accurate, and clearly written laboratory outlines the purpose and method of the experiment, summarizes the work performed, and lists all the results obtained. A Laboratory Report with all data pages will be turned in for each experiment by the end of the class period. Enter all measurements, intermediate, and final results complete with units. Show all work, using dimensional analysis when applicable.

The following general rules should be observed in the preparation of the report:

1. Complete the title page, purpose, method, diagram/list of equipment, chemicals, waste disposal, chemical reactions and mathematical equations sections of the report <u>prior to lab</u>, after a thorough study of the principles and procedures of each experiment from the text in the laboratory manual. These sections (the "pre lab") will be turned in and checked at the start of the laboratory session while the quiz is administered. They will be returned after the quiz.

2. Record and include all measurements and results directly in the report, as they are obtained during the performance of the experiment.

3. In recording quantitative data, make sure that the equipment is read properly (see the Appendix), and that the numbers are followed by proper units.

The laboratory report consists of the following sections:

<u>Title Page</u>:   Fill in the complete title of the experiment as it appears in the manual, the date of the experiment and your name.

<u>Purpose</u>:   Briefly state the purpose of the experiment.

<u>Method</u>:   <u>Summarize</u> the general procedure to be followed in the experiment. <u>Do not</u> copy the printed instructions directly from the manual unless instructed. Instead, write a summary of the method in your own words, omitting minute details but including all the essentials so that another person could repeat the experiment using your report.

<u>Diagram</u>:   A <u>neat</u>, well-labeled diagram of the apparatus, drawn to approximate proportions, should be included whenever applicable. Drawings of often–used items (test tubes, balances, beakers, etc.) are not required. If no drawing is included, list the equipment to be used.

<u>Chemicals</u>:   List all the chemicals used in the experiment.

<u>Waste Disposal</u>:   List the method of all waste disposed in the experiment.

<u>Chemical Reactions</u>:   List all the chemical reactions used in the experiment.

<u>Mathematical Equations</u>:   List the mathematical expressions used in the experiment.

<u>Conclusion</u>:   At the conclusion of the experiment, write a paragraph, in complete sentences using the third person ("it was determined"). First, summarize the experiment briefly (2-3 sentences). Next, state the qualitative and quantitative results obtained through observation and calculation during the course of the experiment. Include units. Not all data collected may be included; relevant data will have to be determined.

<u>Error Analysis</u>:   This will be completed as directed by the instructor.

310

APPENDIX 3: THE CHEMISTRY LABORATORY REPORT

GENERAL CHEMISTRY

LABORATORY REPORT

Experiment Number: ___33___

Title: Mass of a Reactive Metal

Name: __________   Lab Section: __________

Partner: __________   Date: __________

Grade: __________

Front of Report

Purpose: To determine the mass of a reactive metal without the use of a balance.

Method:
1. The apparatus will be set up as shown in Figure 33.1. Add water to the collection bottle and invert into a pan of water.
2. A 5 cm strip of magnesium metal will be sanded to remove the oxide coating and rolled into a small, tight coil.
3. 20 mL of HCl is added to the test tube of the apparatus and the magnesium sample added to the test tube.
4. The hydrogen gas evolved will displace the water in the collection bottle.
5. When the reaction is complete, mark the level of gas in the collection bottle. Drain and measure with tap water how much gas was generated.
6. Using the chemical equation for the reaction and the ideal gas equation, calculate the mass of the magnesium.

Diagram:

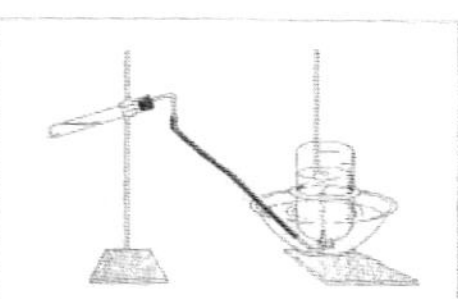

Inside Left

Chemicals:
Magnesium
Hydrochloric acid, 1 M
Hydrogen will be evolved

Waste: All waste can be disposed down the drain.

Chemical Reactions:

$$Mg_{(s)} + 2\ HCl_{(aq)} \rightarrow MgCl_{2\ (aq)} + H_{2\ (g)}$$

Mathematical Equations:

$$P_{H_2} = P_{atm} - P_{water}$$

$$PV = nRT \text{ (to calculate moles of } H_2)$$

$$\text{\# moles } H_2 \times \frac{1 \text{ mole Mg}}{1 \text{ mole } H_2} \times \frac{24.31 \text{ g Mg}}{1 \text{ mole Mg}} = g \text{ Mg}$$

Conclusion:
Hydrochloric acid was reacted with a sample of magnesium in a test tube, resulting in the evolution of hydrogen. The hydrogen was collected over water in a collection bottle. Using the ideal gas equation, the moles of hydrogen were calculated, and using the chemical reaction equation, the mass of magnesium was determined.

Inside Right

Conclusion, continued
519 mL (0.519 L) of hydrogen gas were collected at 22.5°C (295.5 K) over water, and the atmospheric pressure was 0.9934 atm. Dalton's Law of Partial Pressures was used to calculate the pressure of hydrogen (0.9673 atm). Using the ideal gas equation, it was calculated that 0.0207 moles of hydrogen were evolved. Using the balanced chemical equation for this experiment, it was calculated that the mass of the original magnesium sample was 0.504 g. As a general trend, the more the sample of magnesium weighs, the greater the volume of hydrogen gas evolved.

Error Analysis:
There were many sources of error. First, if the system was not air tight, some hydrogen gas could escape from the apparatus. This would lead to a lower volume of hydrogen collected, thus lower moles calculated, and the calculated mass of the magnesium would be lower than expected. If the magnesium oxide was not fully sanded off of the strip of magnesium, the MgO would also be producing hydrogen gas, and the mass of Mg calculated would be higher than it should be. A third source of error would be if the hydrogen gas pressure was not corrected for water vapor pressure, then the pressure would be falsely high giving too high of a mass of Mg. Also, if the bottle was not totally upright or the volume marking was not exact, this would cause faulty hydrogen measurements. It is important to bundle the magnesium tightly and to secure the system immediately after adding the magnesium.

Back of Report

311

LABORATORY REPORT DATA SHEET

R value used: _____ 0.0821 L atm/mol K _____________

	Value	Value (to match R)
Temperature	22.5 °C	295.5 K
Atmospheric pressure	755.0 mm Hg	0.9934 atm
Partial pressure of water at experiment temperature (record/interpolate this value in mm Hg from the Appendix)	0.0261 atm	
Partial pressure of hydrogen at experiment temperature		0.9673 atm
Volume of hydrogen collected	519 mL	0.519 L

Moles of hydrogen generated:

$$n_{H_2} = \frac{PV}{RT} = \frac{(0.9673 \text{ atm})(0.519 \text{ L})}{(0.0821 \frac{\text{L atm}}{\text{mole K}})(295.5 \text{ K})} = 0.0207 \text{ moles } H_2$$

Mass of magnesium reacted:

$$0.0207 \text{ moles } H_2 \times \frac{1 \text{ mole Mg}}{1 \text{ mole } H_2} \times \frac{24.31 \text{ g Mg}}{1 \text{ mole Mg}} = 0.504 \text{ g Mg}$$

Inserted Data Page

APPENDIX 4: PROPER READING
OF LABORATORY EQUIPMENT

To correctly read a piece of equipment, first determine the magnitude of the smallest division (markings). Then estimate the reading in the next smallest order of magnitude. For example, if a thermometer is marked in one-degree increments, the reading would be estimated to the tenth of a degree. If it is marked in tenth of a degree increments, the reading would be estimated to the hundredth of a degree. Read at the bottom (or top, if curved downward) of the meniscus, if applicable.

Equipment	Smallest Division	Read to nearest	Examples
Centimeter ruler	0.1 cm	0.01 cm	9.67 cm; 8.00 cm
10 mL graduated cylinder	1 mL	0.1 mL	3.8 mL
25 mL graduated cylinder	1 mL	0.1 mL	6.7 mL; 15.6 mL
100 mL graduated cylinder	10 mL	1 mL	73 mL; 80. mL
1000 mL graduated cylinder	10 mL	1 mL	855 mL; 840 mL
Barometer	1 mm Hg	0.1 mm Hg	754.9 mm Hg
Thermometer (1° graduation)	1 °C	0.1 °C	76.5 °C; 120.0 °C
Thermometer (0.1° graduation)	0.1 °C	0.01 °C	76.55 °C; 120.03 °C
50 mL buret	0.1 mL	0.01 mL	33.23 mL; 0.00 mL

Other Equipment:

10 mL pipet	On the red line: calibrated at 10.00 mL
Analytical balance	Read ALL digits of the balance (e.g.: 32.4179 g)

312

APPENDIX 5: SIGNIFICANT FIGURES
*** The Significant Figure Policy is in the Font of the Manual ***

In making chemical calculations involving measured quantities, numbers usually represent approximations rather than exact quantities. To indicate the reliability of these approximations, the concept of significant figures is used. By convention, the significant figures in a number are those that are known with certainty and the first doubtful figure. For example, in measuring the volume of water contained in a cylinder marked in milliliters, we note that the meniscus is between the 14 and 15 mL markings and seems to be about halfway between them. We record the volume as 14.5 mL. In doing so, we are using three significant figures; the 1 and the 4 are known with certainty, but the 5 is estimated. Any more digits to the right of the 5 would have no significance and should not be recorded.

A zero may or may not be significant. When used solely to locate the decimal point (leading zeros), as in 0.0054 (5.4×10^{-3}) zeros are not significant. Zeros between nonzero digits in a number (captive zeros) such as 5004.3 are significant. Zeros occurring at the end of a number (trailing zeros) are significant, <u>only if a decimal point is present.</u> The number 96,500. has five significant figures. If written 96,500, it has three significant figures. This is correctly represented as 9.65×10^{4}. It is often helpful to write the number in scientific notation to determine significant figures.

	Number of Significant Figures	Comments
0.065	2	Write as 6.5×10^{-2} to help decide SF
0.00065	2	Write as 6.5×10^{-4} to help decide SF
650.00	5	Write as 6.5000×10^{2} to help decide SF
1.0065	5	
1.00065	6	
100650	5	The last zero is not significant
100650.	6	The last zero is significant due to the decimal

<u>The Use of Significant Figures in Laboratory Reports</u>:
When deciding how many figures are significant in the final answer, all of the numbers which entered into the final calculation must be examined. The mathematical operation in which numbers are involved makes a difference in the way in which digits are determined significant.

<u>Addition and Subtraction</u>
When adding or subtracting numerical quantities, the quantity with the fewest digits after the decimal point will limit the number which can be written for the answer.

<u>Example</u>:   Add   5.821
               182.6
       +   <u>0.327</u>
              188.748
       =   188.7 written with only one digit after the decimal point because of 182.6

APPENDIX 5: SIGNIFICANT FIGURES

<u>Multiplication and Division</u>
When a result is calculated by multiplying or dividing two or more numerical quantities, the result cannot be any more precise than the least precise of the quantities involved in it determination. A simple rule is to write the answer with as many significant figures as the least number in any quantity involved in the calculation of the answer.

<u>Example</u>:

$$V = \frac{(10.0)(1.455)}{0.22} = 66$$

The answer is limited to two significant figures by 0.22.

<u>Summary</u>
1. Count digits in the number disregarding the decimal point.
    Examples: 5.18 - 3 sf; 51.8 - 3 sf; 581 - 3 sf

2. Zeros are significant only if preceded by a number other than zero.
    Examples: 0.0040 - 2 sf; 0.400 - 3 sf; 400.0 - 4 sf

3. Significant figures in final calculations are dictated by the mathematical operation. Addition and subtraction answers have digits to the right of the decimal limited by the number with the fewest numbers to the right of the decimal. Multiplication and division answers are determined by the number with the fewest significant figures.

4. Keep one more significant figure than allowed in final answer through all intermediate calculations.

5. Exact numbers obtained from conversion factors, definitions or by counting numbers of objects have an infinite number of significant figures.

(NOTE: The same rules are used in "rounding off" numbers as in algebra: "round off" to the next higher digit if number is 5 or more than 5.)

Notes on significant figures for K_a and pH in acid-base problems (Chem1114)

$$K_a = 1.8 \times 10^{-5} \qquad pH \; = \; 2.66$$

coefficient   base   exponent   characteristic   mantissa

1. The number of significant figures in a K_a or K_b value is the number of significant figures in the <u>coefficient</u>. The K_a value above has two significant figures.

2. The number of significant figures in a pH value is the number of significant figures in the <u>mantissa</u>. The pH value above has two significant figures.

3. The molarity determined from using one of the above numbers would also limited to two significant figures due to multiplication and division rules.

APPENDIX 6: CONSTANTS AND CONVERSION FACTORS

<u>R</u>

$$0.0821 \; \frac{L \; atm}{mole \; K} \qquad 62.4 \; \frac{L \; torr}{mole \; K} \qquad 8.314 \; \frac{J}{mole \; K} \qquad 8.314 \; \frac{m^3 \; Pa}{mole \; K}$$

<u>Pressure</u>

$$1 \; atm \; = \; 101{,}325 \; Pa \; = \; 29.92 \; inches \; Hg \; = \; 760 \; mm \; Hg \; = \; 760 \; torr$$

<u>Temperature</u>

$$K = {}^{\circ}C \; + \; 273.15 \qquad\qquad {}^{\circ}C \; = ({}^{\circ}F \; - \; 32)\frac{5}{9}$$

<u>Mass</u>

$$1 \; kg \; = \; 1000 \; g$$

<u>Volume</u>

$$1000 \; mL \; = \; 1 \; L$$
$$1 \; cm^3 = 1 \; mL$$

<u>Other</u>

$$\text{Avogadro's number: } 6.022 \times 10^{23}$$
$$\text{Faraday Constant (F): } 96485 \; C/mol \; e^- \approx 96500 \; C/mol \; e^-$$

APPENDIX 7: ACTIVITY SERIES OF METALS

Li
K
Ba
Ca
Na
Mg
Al
Zn
Cr
Fe
Cd
Co
Ni
Sn
Pb
H$_2$
Cu
Ag
Hg
Pt
Au

Reducing Strength Increases (Reactivity Increases)

APPENDIX 8: BALANCING REDOX REACTIONS

<u>Ion-Electron Method for Balancing Redox Equations</u>

1. Write the unbalanced ionic equations for the two half-reactions.
   Do steps 2 – 5 for one half reaction at a time.

   2. Balance atoms other than H and O in the ½ reaction.
   3. Balance O with H_2O.
   4. Balance H with H^+.
   5. Balance charge with appropriate number of electrons.

6. Multiply balanced half-reactions by appropriate coefficients so that the number of electrons are equal and will cancel each other out.

7. Add the half-reactions together.

8. If in acidic solution, continue with step 9. If in basic solution, add an equal number of OH^- groups to <u>both</u> sides of the equation to neutralize all of the H^+. (Note: if OH^- and H^+ are on the same side, combine to make water.)

9. Cancel species that appear on both sides to get the balanced net ionic equation.

10. If necessary, add spectator ions to get the balanced molecular equation.

11. Check the final balance (atoms and charges)!

APPENDIX 9: VAPOR PRESSURE OF WATER

°C	atm	mm Hg or torr
10.0	0.0121	9.2
11.0	0.0129	9.8
12.0	0.0138	10.5
13.0	0.0147	11.2
14.0	0.0158	12.0
15.0	0.0168	12.8
16.0	0.0179	13.6
17.0	0.0191	14.5
18.0	0.0204	15.5
19.0	0.0218	16.6
20.0	0.0230	17.5
21.0	0.0246	18.7
22.0	0.0261	19.8
23.0	0.0278	21.1
24.0	0.0295	22.4
25.0	0.0313	23.8
26.0	0.0332	25.2
27.0	0.0351	26.7
28.0	0.0372	28.3
29.0	0.0395	30.0
30.0	0.0418	31.8
31.0	0.0443	33.7
32.0	0.0470	35.7

> For values in between those listed in the table, assume these values are essentially linear and interpolate.

To interpolate, use the point-slope formula (°C is the y, and the pressure is the x). Find the slope from the two sets of known points. Use the point slope formula, with the higher temperature as y, the desired temperature for y_1, and the higher pressure for x. Solve for x_1.

$$y - y_1 = m(x - x_1)$$

Example: Find the vapor pressure of water in mm Hg at 27.3 °C.

$$28.0 - 27.3 = \left(\frac{28.0 - 27.0}{28.3 - 26.7}\right)(28.3 - x_1) \qquad x_1 = 27.2 \text{ mm Hg}$$

APPENDIX 10: SELECTED POLYATOMIC IONS

+1	−1	−2	−3
NH_4^+ (ammonium)	CN^- (cyanide)	O_2^{-2} (peroxide)	
H_3O^+ (hydronium)	OCN^- (cyanate)	$C_2O_4^{-2}$ (oxalate)	
Hg_2^{2+} (mercury (I))	SCN^- (thiocyanate)		
	OH^- (hydroxide)	CrO_4^{-2} (chromate)	
	$C_2H_3O_2^-$ (acetate)	$Cr_2O_7^{-2}$ (dichromate)	
	MnO_4^- (permanganate)		
	NO_3^- (nitrate)		
	NO_2^- (nitrite)		
	$H_2PO_4^-$ (dihydrogen phosphate)	HPO_4^{-2} (hydrogen phosphate)	PO_4^{-3} (phosphate)
	HCO_3^- (hydrogen carbonate or bicarbonate)	CO_3^{-2} (carbonate)	AsO_4^{-3} (arsenate)
	HSO_4^- (hydrogen sulfate	SO_4^{-2} (sulfate)	
	HSO_3^- (hydrogen sulfite)	SO_3^{-2} (sulfite)	
	ClO_4^- (perchlorate)	$S_2O_3^{-2}$ (thiosulfate)	
	ClO_3^- (chlorate)		
	ClO_2^- (chlorite)		
	ClO^- (hypochlorite)		
	+ bromo, iodo relatives		

Courtesy of Dr. Patty Wisian-Neilson

APPENDIX 11: SAMPLE GRADING RUBRIC (70 POINT LAB)

Section	Points	Description (numbers in parentheses are point breakdowns)
Purpose	1	Clear, complete sentence(s)
Method	1	Clear, succinct – take off points if extremely abbreviated
Diagram	1	Or list equipment
Chemicals	3	$Fe(NO_3)_3$ (Iron (III) nitrate)
		KSCN (Potassium thiocyanate)
		HNO_3 (Nitric acid)
Waste	1	All waste can be disposed down the drain
Reactions	1	$Fe^{3+} + SCN^- \rightarrow Fe(SCN)_x^y$
Equations	3	$A = \varepsilon b C$

$$K_f = \frac{[Fe(SCN)_x^y]}{[Fe^{3+}]^z[SCN^-]^x}$$

$$[\ ]_0 = \frac{(Molarity)(Volume\ of\ species)}{total\ volume}$$

Data			Data Pages	Included (All values with correct number of digits or SF)
(27)	2			Correct ICE table and equations
	15	Results		Correct $[Fe(SCN)_x^y]$ with units 1.82×10^{-4} M (2)
				Correct ε with units ($\sim$6000 M^{-1}) (2)
				Correct complex formula, $Fe(SCN)^{2+}$ (5)
				K_f: correct order of magnitude (10^2) (5) ($\sim$150)
				K_f has no units; if units are included: -1
	10	Graph		Title (2)
				x axis label (2)
				y axis label (2)
				correct shape (2)
				goes through (0,0) and (1,0) (2)
Conclusion	14	Contains		ε (3)
				Formula of the complex (4)
				K_f value (4)
				Chemical reaction equation (3)

Error Analysis   18   Answers the following questions:

1.   State Beer's Law in words and with an equation.

(2)   Words: The absorbance is equal to the product of the concentration times
                                    the path length times the molar absorptivity

(1)   Equation: $A = \varepsilon b C$

2 (a). (3) When determining ε, if 2.00×10^{-3} M $Fe(NO_3)_3$ was mistakenly used instead of 0.200 M $Fe(NO_3)_3$, how
            would the absorbance be affected? Explain.
                        The absorbance would be lower, because the concentration of $Fe(NO_3)_3$ would be lower.

  (b). (3) How would this affect the experimentally determined value of ε?
                        The ε would decrease because the absorption using the lower concentration $Fe(NO_3)_3$ would be lower.

3. (3)   In any reaction, what is a typical <u>minimum</u> value of K_f? ______Greater than 100___

4. (3)   What does a large K_f mean in terms of equilibrium? ____Eq lies to the right/products___

5. (3)   How do scratches or fingerprints on the cuvette affect the absorbance? ___Increase the absorbance____

APPENDIX 12: DIGITAL pH METER INSTRUCTIONS

A pH meter provides a precise method to measure the pH of aqueous solutions. A pH meter is an electrical device that measures the difference in electrical potential, i.e. voltage, developed between two electrodes which are immersed in a solution. One electrode is a reference electrode and the other a measuring glass electrode, which are often combined into one unit called a combination electrode. The electrical potential of the reference electrode is fixed while the potential of the glass electrode varies with the H^+ concentration in the solution being measured. The voltage difference between the two electrodes is electronically amplified and converted to pH units, which are read directly on the meter. The meter must be calibrated with known pH solutions before each use.

Standardize the meter and electrode using at least two buffers with pH values bracketing the expected pH of the samples (usually pH 4 and pH 7 buffers are used). The default temperature is 25°C.

1. Press **setup** on the digital pH meter until "**clear**" appears in the lower part of the display. Press **enter**. If **clear** does not appear but "2 4 7 10 12" appears, press enter and continue with step 3.

2. Press **setup** on the digital pH meter until the lower left corner of the screen displays the numbers 2 4 7 10 12 (not 1, 3, 6, 8). Press **enter**.

3. Unscrew the pH probe from the storage container (if applicable). Leave this top on during the calibration and for all subsequent readings. Normalize the electrode with a small amount of the first buffer solution by pouring some of the buffer solution over the electrode tip. Gently shake off any excess.

4. Immerse the electrode in the first buffer solution. Stir gently with the electrode for a few seconds. Allow the electrode to reach a stable value (an **S** will show on the left side of the screen).

5. Press **standardize**, allow time for the electrode to stabilize (it will cycle through some readings, then 100.0, then back to the pH of the buffer).

6. When the "S" appears again and the first buffer number appears in the lower left, press enter. Normalize the electrode with a small amount of the second buffer by pouring a small amount of the second buffer over the electrode tip. Gently shake off any excess solution and then place the electrode in the second buffer solution. Stir briefly, allow time for the electrode to stabilize, and press **standardize**. When the signal is stable (an "S" appears in the display), the second buffer number will appear on the display. Press enter.

7. Normalize with a third buffer if directed.

8. When buffers are correctly entered, the numbers 4 and 7 (and the third buffer, if done) will display in the bottom left corner of the screen. Standardization is complete. The pH meter is now ready to test solutions. Do not press any buttons for the duration of the experiment. Dispose of buffers down the drain.

Note: Normalize the electrode with deionized water and then gently shake off the excess before testing solutions. During titrations, swirl the solution with the electrode, then let the electrode sit still and wait for the values to stabilize, then record the pH.

<u>Always keep the pH electrode immersed and upright when not in use.</u>
<u>Notify the instructor if more 3 M KCl solution needs to be added to the storage container.</u>

APPENDIX 13: ERROR ANALYSIS

Most of the experiments in this Manual involve measurement of physical quantities: mass of a substance, volume of a gas, temperature, volume of a solution, etc. All measurements are subject to error and these errors must be taken into account in drawing any conclusions based on experimental results.

In judging the results of an experiment, there are two questions that must be considered - how precise the measurements are, and how accurate they are. The precision refers to the reproducibility of a number of repeated measurements. The accuracy refers to how well the values agree with the true value. The accuracy is measured by a quantity known as "the experimental error" which is defined as:

$$\% \text{ error} = \frac{\text{experimental value - theoretical value}}{\text{theoretical value}} \times 100$$

A positive % error means the experimental value is larger than the theoretical value. A negative % error means the experimental value is smaller than the theoretical value. It gives a magnitude of the error, but does not indicate what type (systematic or random) error it is.

The error analysis section of the laboratory report will be different for each lab as directed by the instructor. When writing errors, it is important to note the error, what parameter it affects, and the direction of the error. An example is given as follows:

> If the acetic acid was overtitrated with NaOH, the volume of the added NaOH would be too great. The calculated molarity of the acetic acid would appear too low.

Errors not allowed in the Error Analysis section for any 1113/1114 lab:
1. The buret or pipet was read incorrectly.
2. The substance was weighed wrong.
3. The buret or pipet was not normalized properly.
4. The significant figures were not recorded properly.
5. The wrong substance was added.
6. The numbers were transposed.

APPENDIX 14: EQUIPMENT

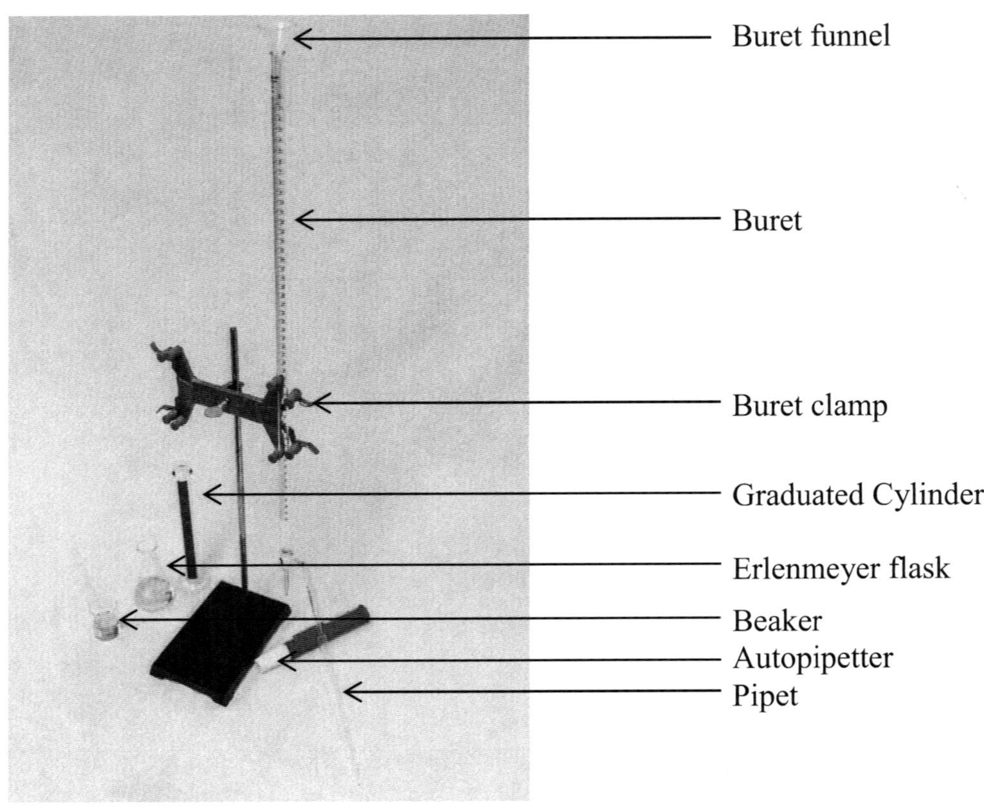

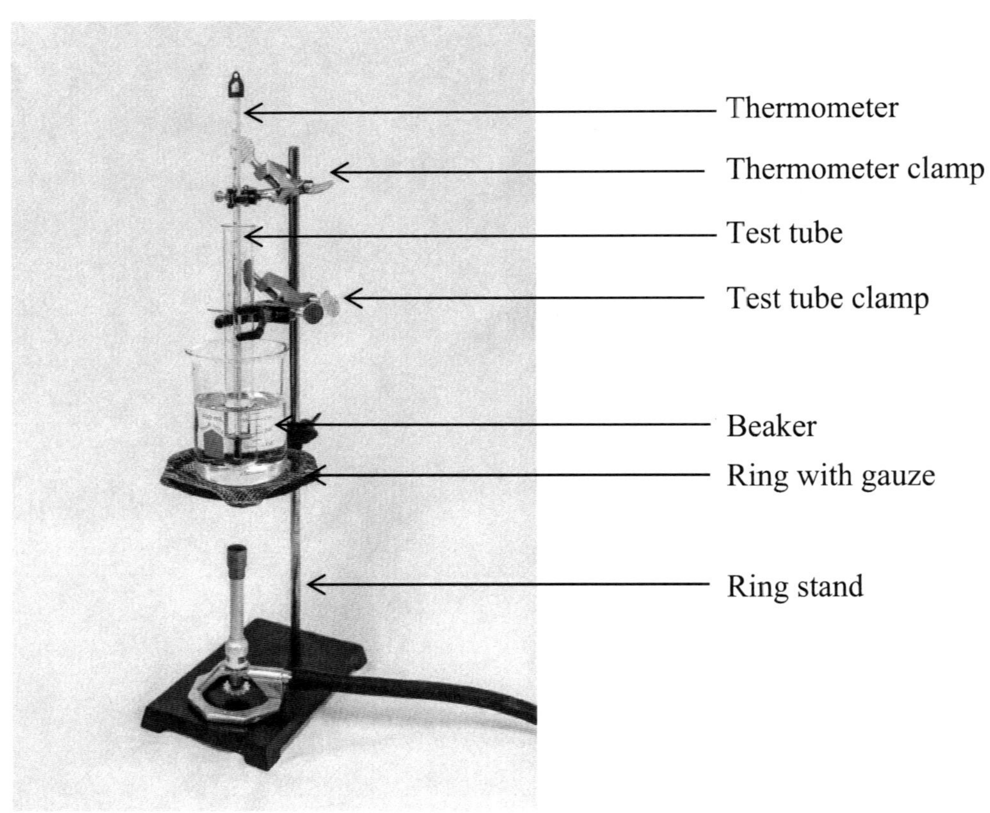

APPENDIX 14: EQUIPMENT

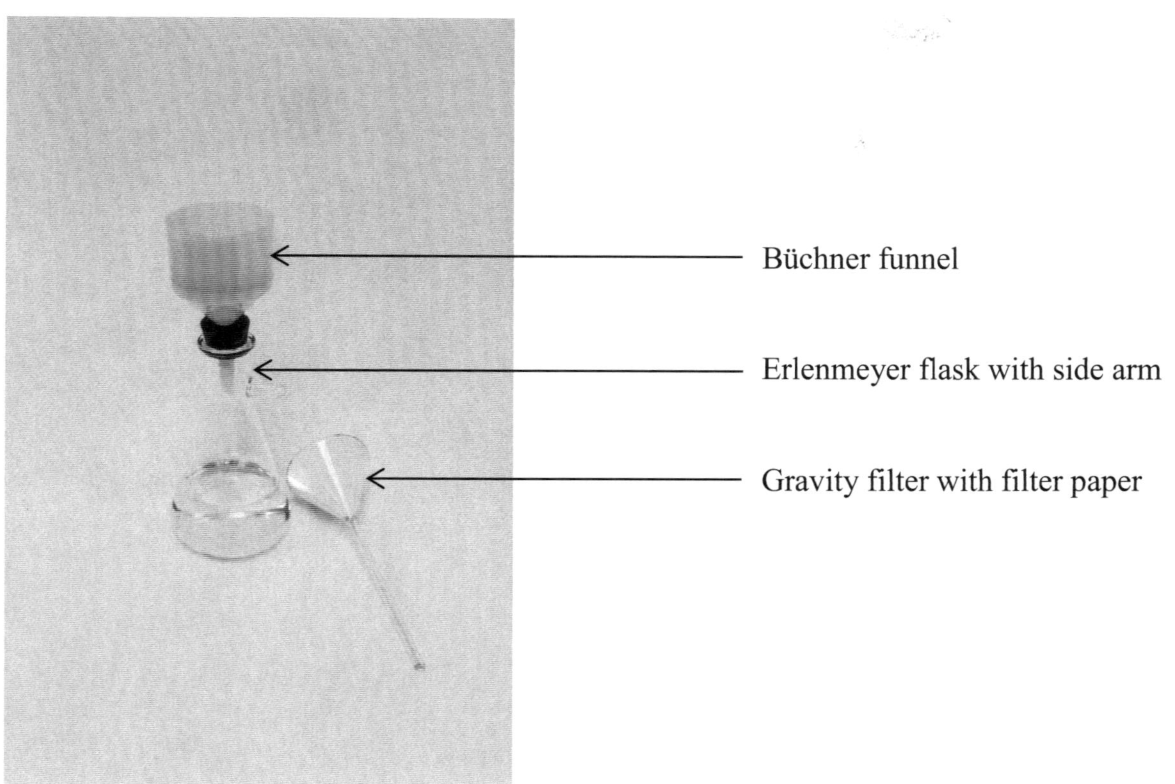

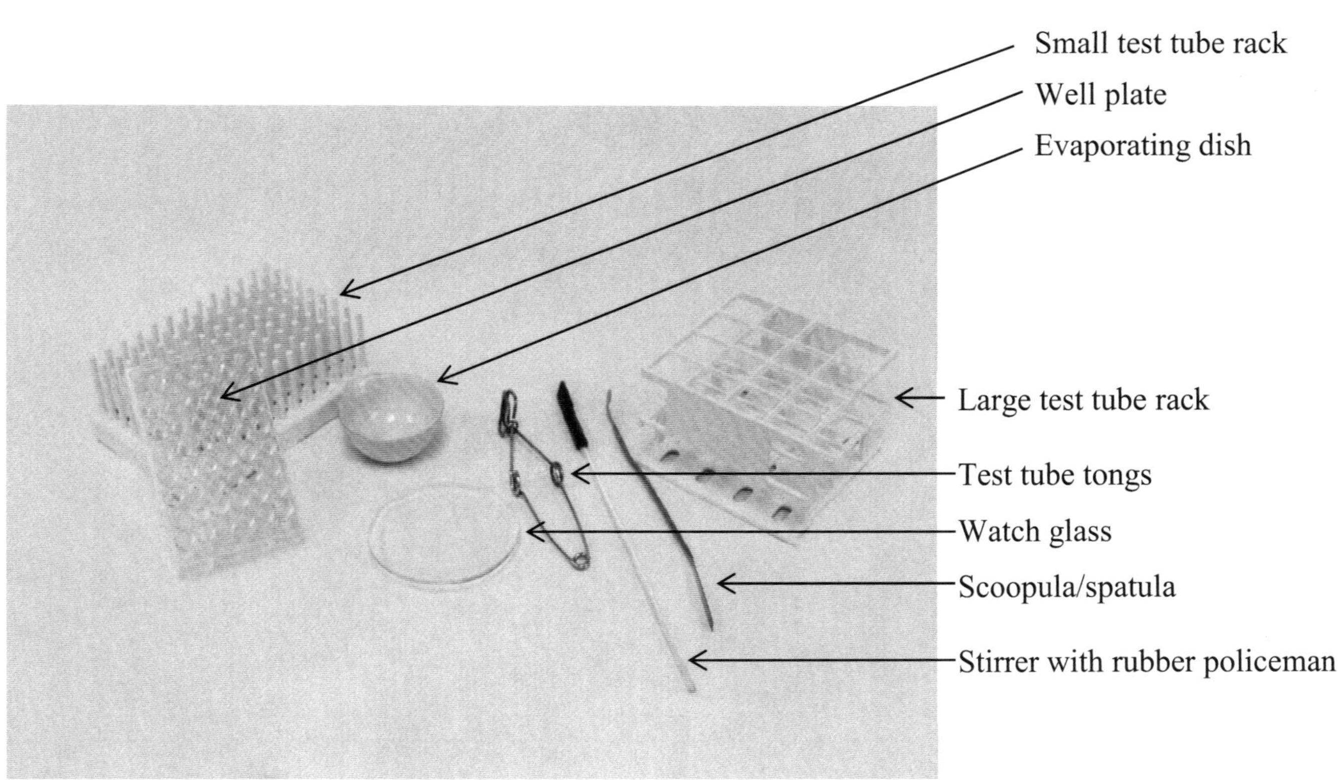

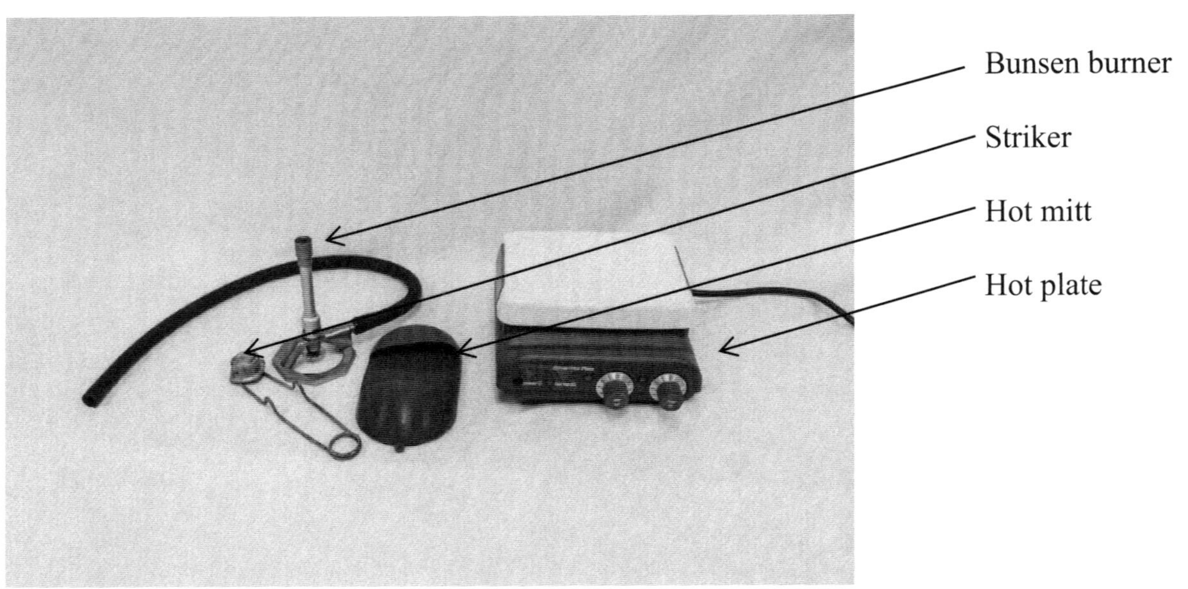
Bunsen burner
Striker
Hot mitt
Hot plate

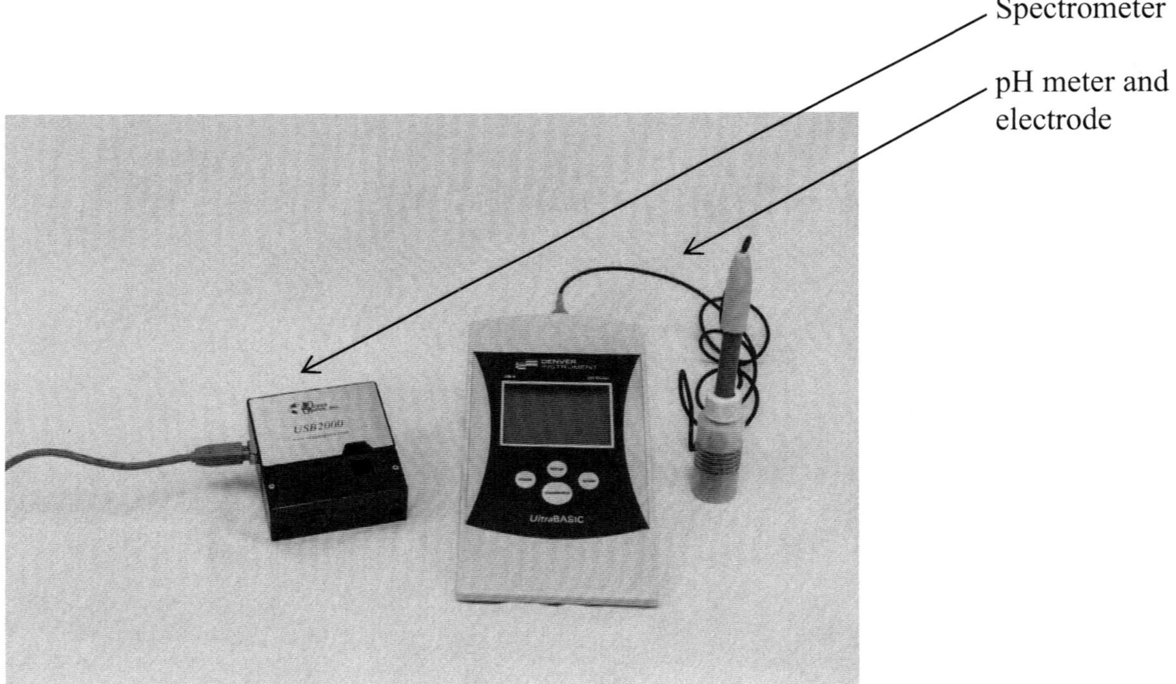
Spectrometer
pH meter and
electrode

APPENDIX 15: ANSWERS TO SELECTED HOMEWORK

Experiment 1: Density of Liquids and Solids

3a.	18.10 g water	3f.	2.17 %
3b.	15.63 g methanol	3g.	926.9 kg/m^3
3c.	53.66 %	4.	0.8332 g/mL
3d.	46.34 %	5.	7.83 g/mL
3e.	0.9269 g/mL	6b.	29.00 %

Experiment 2: The Copper Cycle

2a. Nitric acid

2b. Oxides of nitrogen (NO, NO_2)

2c. It reacts with the toxic gases, removing them from the solution

2d. No brown gas on the test tube walls, solution is bright blue

3a. Zinc

3b. $Zn_{(s)} + 2\ HCl_{(aq)} \rightarrow ZnCl_{2\ (aq)} + H_{2\ (g)}$

3c. The solution is colorless

4a. 0 and 100

4b. 82.293 %

5a. 10.9 g HNO_3

5b. 3.98 g NO_2

5c. 1.5042 g Zn

Experiment 3: Stoichiometric Calculations and Graphing

Dilution

1. 0.18 M

2. 0.55 %

3. Cannot be used. Not a dilution.

4. 20.0 mL

5. 0.694 M

Stoichiometric Calculations

1a. 10.12 g N_2

1b. 8.777 x 10^{-3} m^3

2a. 3.06 g $CaCl_2$

2b. 4.1220 Na_3PO_4

2c. 3.9381 g NaCl

3b. 10.688 g Fe

3c. 9.7555 g Al_2O_3

3d. 15.280 g Fe_2O_3

Graphing, Extrapolation, and Scaling

1. Top left graph: Volume vs Pressure; x axis label: Pressure (atm); y axis label: Volume (mL)

2. 0.83 g Cu/50 mL

3. Volume (0-32 mL) 4, 0.4 or 5, 0.5

4. pH 2, 0.2

Experiment 4: Limiting Reagents

1a. $CaCl_{2\ (aq)} + 2\ KOH_{(aq)} \rightarrow Ca(OH)_{2\ (s)} + 2\ KCl_{(aq)}$

1b. Net ionic: $Ca^{2+}_{\ (aq)} + 2\ OH^-_{\ (aq)} \rightarrow Ca(OH)_{2\ (s)}$

1c. 11.92 g $Ca(OH)_2$

1d. 0.13 g excess $CaCl_2$

1e. 88.59 %

2c. 13.03 g $Cu_3(PO_4)_2$

2d. 1.47 g Na_3PO_4

2e. 76.4 %

3c. 17.29 g PbI_2

3d. 10.77 g $Pb(NO_3)_2$

3e. 0.06503 M

4b. 76.5 %

4c. 37 kg acetic anhydride

Experiment 5: Polyatomic Ions, Solubility Rules, Net Ionic Equations, and Oxidation States
1.	Insoluble halide: Ag^+, Hg_2^{2+}, Pb^{2+} with either Cl^-, Br^-, or I^- (Example: AgCl)
	Insoluble nitrate: none
2b.	One possibility: $Mg(OH)_2$, $MgSO_4$
3.	magnesium sulfate + calcium chloride
	Molecular: $MgSO_{4\ (aq)} + CaCl_{2\ (aq)} \rightarrow MgCl_{2\ (aq)} + CaSO_{4\ (s)}$
	Ionic: $Mg^{2+}_{\ (aq)} + SO_4^{2-}{}_{(aq)} + Ca^{2+}_{\ (aq)} + 2\ Cl^-_{\ (aq)} \rightarrow Mg^{2+}_{\ (aq)} + 2\ Cl^-_{\ (aq)} + CaSO_{4\ (s)}$
	Net Ionic: $Ca^{2+}_{\ (aq)} + SO_4^{2-}{}_{(aq)} \rightarrow CaSO_{4\ (s)}$
5.	$KClO_4$: Cl = +7
	$Bi_2(CrO_4)_3$ = +3

Experiment 6: Acid-Base Titrations
1.	0.2650 M NaOH
2a.	0.1807 M NaOH
2b.	292.1 g/mole
2c.	584.3 g/mole
3b.	1608 mL

4a.	0.3864 M acetic acid
4b.	23.21 g/L
4c.	2.32 %
4d.	0.171 L

Experiment 7: Oxidation-Reduction Titrations
1b.	0.03514 M
3a.	0.6146 g H_2O_2
3b.	38.05 %
4.	91.8 mL

5a.	0.2021 M
5b.	218 mL
6b.	261.8 g/mole

Experiment 8: Heats of Reaction
1b.	–2.52 kJ
1c.	–472 kJ/mol
1d.	1.10 %
2.	26.9 °C
3.	7.38 °C

4b.	–8.20 kJ
4c.	–56.2 kJ/mol
4d.	–59.6 kJ/mol
4e.	They do have an impact, but it is considered negligible

Experiment 9: Gas Laws
3a.	$Mg_{(s)} + 2\ HCl_{\ (aq)} \rightarrow MgCl_{2\ (aq)} + H_{2\ (g)}$
3b.	0.9455 L
4c.	$Na_2SO_{3(s)} + 2\ HCl_{\ (aq)} \rightarrow NaCl_{\ (aq)} + H_2O_{(l)} + SO_{2(g)})$

4b.	2.51 g/L
4c.	–4.6%
4d	2.51×10^{-3} g/cm^3.

Experiment 10: Bonding and Molecular Geometry
2.	Linear, trigonal planar, tetrahedral, trigonal bipyramidal, octahedral
3.	They push atoms closer together, making the angles smaller
6.	3 bonding and 1 lone pair: tetrahedral arrangement, trigonal pyramidal molecular geometry, and sp^3 hybridization
7.	C–Br bond is polar, while the molecule is non polar
8.	To make each bond in a molecule identical

Experiment 11: Molar Mass Determination by Freezing Point Depression
1a.   5.4 °C, 3.4 °C, 1.7 °C
1b.   150 g/mole
1c.   150 g/mole
2.   Solvent may not be pure; thermometer may not be accurate
3.   Research this in any General Chemistry textbook; it's usually under the discussion
        of Boiling Point Elevation
4.   16,500 g/mol
5.   −26.38 °C

Experiment 12: The Rate of Chemical Reactions
1.   Run 1: $[H_2SeO_3]$ = 0.0500 M; $[I^-]$ = 0.300 M; $[H^+]$ = 0.200 M; $\Delta[Se]$ = 0.0500 M;
        Rate = 1.67 x 10^{-4} M/s; k = 3.09 M^{-5} s^{-1}
        Order with respect to H_2SeO_3: 1
        Order with respect to I^-: 3
        Order with respect to H^+: 2
2b.   Rate = $k[N_2O_5]$
2c.   k = 0.0277 s^{-1}
3c.   k = 0.200 $M^{-1}s^{-1}$

Exeriment 13: Quantitative Analysis of an Alloy

1a.	0.33 g Cu	2b.	0.95
1b.	43.5 % Cu	3c.	NO, NO_2
1c.	0.0349 g Ag	3d.	NaOH
1d.	0.175 g Ag	4a.	Net Ionic: $Ag^+_{(aq)}$ + $SCN^-_{(aq)}$ $\rightarrow$ $AgSCN_{(s)}$
1e.	23.1 % Ag	4b.	Ferric alum
2a.	3.57 mL		

Experiment 14: Le Châtelier's Principle

2a.	Concentration, temperature, pressure, volume	3e.	No effect
2b.	Temperature	3h.	$\rightarrow$, $\downarrow$, $\uparrow$
3a.	$\rightarrow$, $\downarrow$, $\uparrow$	3j.	$\leftarrow$, $\uparrow$, $\downarrow$
3d.	$\rightarrow$, $\downarrow$, $\uparrow$	4b.	No change; solids do not affect the equilibrium
		4e.	K_c = $1/[SO_2]$

Experiment 15: Acids and Bases: pH, Molarity K_a, and K_b
1. $[H^+]$ = 6.84 x 10^{-3} M; pH = 2.17; pOH = 11.83; $[OH^-]$ = 1.48 x 10^{-12} M; K_b = 5.9 x 10^{-11} M
3. $[H^+]$ = 2.40 x 10^{-3} M; M = 0.322 M; pOH = 11.38; $[OH^-]$ = 4.17 x 10^{-12} M; K_b = 5.6 x 10^{-10} M
5.   2.49 %, 1.55 %, 0.745 %. 0.807 %
7. CH_3COOH < C_6H_5COOH < HCOOH; the higher the K_a, the more acidic the acid
9.   0.275 M CH_3COOH < 0.322 M CH_3COOH; the greater the molarity, the stronger the acid
        OR the lower the pH, the stronger the acid
10. $[H^+]$ = 3.02 x 10^{-12} M; pH = 11.52; pOH = 2.48; $[OH^-]$ = 3.29 x 10^{-3} M; K_a = 5.9 x 10^{-10} M
11b-d. 33.7 mL, 16.9 mL, 4.74

Experiment 16: Acids and Bases: Complete Curve Analysis

1a.	Initial pH: 2.10		2b.	5.58×10^{-11}
	Initial $[H^+]$: 7.94×10^{-3} M		2c.	pH = 2.22
	EP mL: 18.50 mL		2d.	pH = 4.04
	EP pH: 8.0		2e.	10.8 mL
	½ EP mL: 9.25 mL		2f.	pH = 8.47
	½ EP pH: 3.3		2g.	pH = 13.30
1b.	$HNO_2 + NaOH \rightarrow NaNO_2 + H_2O$		2h.	pH = 3.77
	$NaNO_2 \rightarrow Na^+ + NO_2^-$		3a.	1.74×10^{-9}
	$NO_2^- + H_2O \rightleftharpoons HNO_2 + OH^-$		3b.	5.75×10^{-6}
1c.	0.143 M HNO_2		3c.	Pyridine
1d.	5.01×10^{-4}		3d.	40.6 mL
1e.	4.41×10^{-4}		3e.	20.3 mL
1f.	The identity of the salt		3f.	1.70×10^{-9}
1g.	pH = 8.16		3g.	pH = 4.93
1h.	The volume and molarity of the base		3h.	pH = 2.98
1i.	pH = 12.52		3i.	pH = 1.32

2a. $2\ HCOOH + Ba(OH)_2 \rightarrow Ba(COOH)_2 + 2\ H_2O$

4. $K_{a1} = 6.3 \times 10^{-3}$; $K_{a2} = 1.0 \times 10^{-7}$; $K_{a3} = 3.2 \times 10^{-12}$; $pH_{1st\ EP}$: 4.60; $pH_{2nd\ EP}$: 9.25

Experiment 17: The Solubility Product Constant for $Cu(IO_3)_2$

1b.	All Cu^{2+} must be converted to the complex ion, so no other species is measured.			
3.	0.0120 M		5b.	2.81×10^{-10} M; 2.89×10^{-8} g/L
4a.	$K_{sp} = [Cr^{3+}][OH^-]^3$		5c.	7.05×10^{-26} M; 7.27×10^{-24} g/L
4b.	3.00×10^{-29}		6.	$K_{sp} = [Ca^{2+}]^3[PO_4^{3-}]^2$
5a.	3.25×10^{-8} M; 3.35×10^{-6} g/L			

Experiment 18: Electrochemistry and Thermodynamics

1. Anode: Mn; $Mn_{(s)} \rightarrow Mn^{2+}_{(aq)} + 2\ e^-$
Cathode: Sn; $Sn^{2+}_{(aq)} + 2\ e^- \rightarrow Sn_{(s)}$
$Mn_{(s)} + Sn^{2+}_{(aq)} \rightarrow Mn^{2+}_{(aq)} + Sn_{(s)}$
$Mn_{(s)} \mid Mn^{2+}_{(aq)}$ (1 M) $\parallel Sn^{2+}_{(aq)}$ (1 M) $\mid Sn_{(s)}$
$E° = 1.04$ V
$\Delta G° = -201$ kJ/mol
$K = 1.71 \times 10^{35}$
+, –, large, products
1.07 V
1.03 V
Increases
Decreases

2. Anode: Zn; $Zn_{(s)} \rightarrow Zn^{2+}_{(aq)} + 2\ e^-$
Cathode: Al; $Al^{3+}_{(aq)} + 3\ e^- \rightarrow Al_{(s)}$
$3\ Zn_{(s)} + 2\ Al^{3+}_{(aq)} \rightarrow 3\ Zn^{2+}_{(aq)} + 2\ Al_{(s)}$
$Zn_{(s)} \mid Zn^{2+}_{(aq)}$ (1 M) $\parallel Al^{3+}_{(aq)}$ (1 M) $\mid Al_{(s)}$
$E° = -0.90$ V
$\Delta G° = 521$ kJ/mol
$K = 4.7 \times 10^{-92}$
–, +, small, reactants
– 0.92 V
– 0.89 V
Decreases
Increases

3. $\Delta S = -222$ J/mol K; $\Delta H\ -677$ kJ/mol

Experiment 19: Electrolysis and Faraday's Laws

1a.	$Cu_{(s)} \rightarrow Cu^{2+}_{(aq)} + 2\ e^-$		2a.	0.3184 A
1b.	$Cu^{2+}_{(aq)} + 2\ e^- \rightarrow Cu_{(s)}$		2b.	21.6 mm Hg or 0.0285 atm
1c.	3600 C		2c.	0.301 L
1d.	0.037 mole e^-		2d.	–7.0 %
1e.	0.037 F			

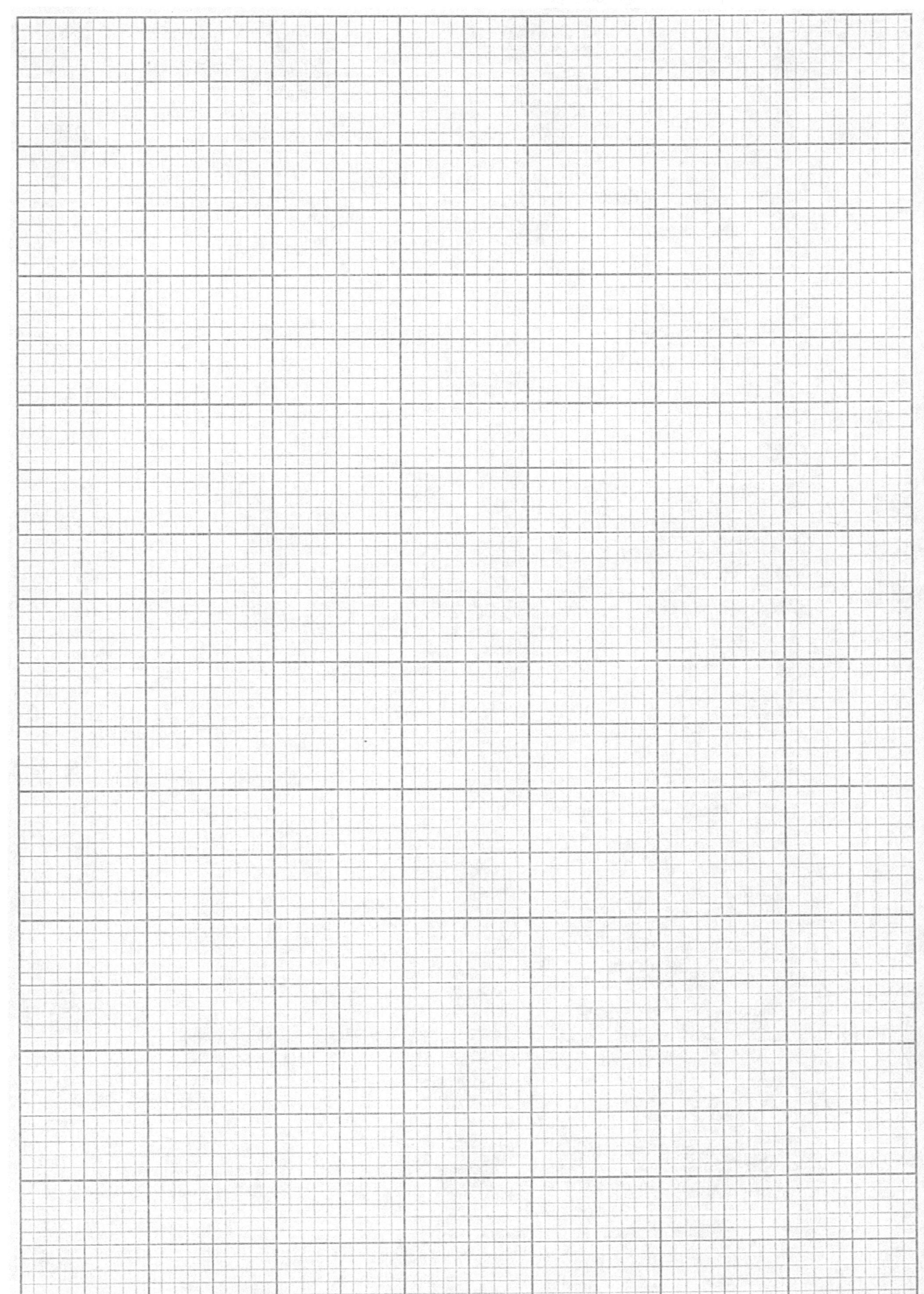